Earth Issues Reader

for

Understanding Earth

Third Edition

Frank Press and Raymond Siever

and

Environmental Geology

Dorothy Merritts, Andrew De Wet,
and Kirsten Menking

W. H. Freeman and Company
New York

For information about subscribing to the *CQ Researcher,* contact Maureen Whelan (mwhelan@cq.com or 800-638-1710, x501); for other inquiries about classroom use of CQ publications, contact James R. Headley (jheadley@cq.com or 202-887-8608).

ISBN: 0-7167-4370-1

© 2001 by W. H. Freeman and Company

No part of this book may be reproduced by any mechanical, photographic, or electronic process, or in the form of a phonographic recording, nor may it be stored in a retrieval system, transmitted, or otherwise copied for public or private use, without written permission from the publisher.

Printed in the United States of America

Second printing, 2001

Contents

The Economics of Recycling 1
Is it worth the effort?

Coastal Development 26
Does it put precious lands at risk?

The Politics of Energy 54
How should Congress handle U.S. energy problems?

Saving Open Spaces 80
Are land-conservation efforts in the public interest?

Oil Production in the 21st Century 106
When will the world run out of oil?

Americans generate over 200 million tons of solid waste every year. From empty toothpaste tubes and soda bottles to refrigerators and cars, virtually all products used in the United States are eventually thrown away. Historically, much of the waste has ended up in landfills, though a portion has been burned. In recent years, federal, state, and local initiatives have encouraged recycling, and 27% of the nation's solid waste stream was recycled by 1998. The most successfully recycled materials include cardboard, 64% of which was recycled in 1995; lead-acid batteries (96%); aluminum cans (63%); and major appliances (61%). Other products such as plastic milk containers and soda bottles, magazines, and glass have lagged behind, as these recyclables have had to compete in the market with cheaper virgin materials. Competition causes market instability for recyclables, a fact that has led to the criticism that recycling is not economically viable. Furthermore, some critics question the necessity of recycling in the first place. Whereas some local communities have run out of landfill space, the conservative policy think tank the Cato Institute has calculated that all of the waste generated by the United States in the next 1000 years could be placed in a space 1000 feet deep and 30 square miles in area. Given that we have plenty of land on which to store our waste and that recycling sometimes costs more than use of virgin materials, a representative of the Institute argues that we should simply landfill our waste.

What critics of recycling fail to take into consideration is that landfilling and use of virgin materials carry environmental consequences that are not reflected in the prices of consumer products. Rainwater percolating through landfills dissolves toxic metals and chemicals and transports them through the groundwater system where they may contaminate wells used for drinking water (see Merritts, De Wet, & Menking, pp. 250–256, for a discussion on groundwater pollution; Press & Siever, pp. 272–274). Mining ore minerals exposes metal sulfides to rain and groundwater, which creates toxic and corrosive runoff that flows into streams and lakes, killing aquatic life (see Merritts, De Wet, & Menking, pp. 132–133, 168, for a discussion on acid mine drainage). Use of virgin materials helps to deplete nonrenewable natural resources (see Merritts, De Wet, & Menking, pp. 17–19, for a discussion on different types of resources—nonrenewable being one type; Press & Siever, p. 508). For example, plastics are manufactured from oil, and by not recycling these materials, we hasten the time when the world will run out of this important resource. Were the full costs of landfilling and use of virgin materials reflected in the price of consumer goods rather than in income taxes used for cleaning up toxic waste sites and for military activities to ensure a steady supply of cheap oil, recycling would prove to be far more cost effective than it presently appears to be. One industrialized nation that has made tremendous strides toward recycling its solid waste stream is Germany, which has instituted a program that requires manufacturers to take back and reprocess their goods and packaging after consumer use. Recycling has become a multibillion-dollar industry in that nation, and manufacturers have learned how to reduce the amount of packaging they use and how to create biodegradable and less toxic materials. Other countries in the European Union are taking steps to follow the German model, but some wonder whether such a system could work in the United States, where landfill space is plentiful and where many citizens resent government programs.

The Economics of Recycling

Is it worth the effort?

In the late 1980s, acting on fears that landfill space was running out, communities across the country began curbside collection of paper, glass, metal and plastic waste. Polls suggest that Americans strongly support recycling, despite the fact that the United States remains the world's leading "throwaway society." But critics say recycling is often a wasted effort, helping consumers' consciences more than the environment or the economy. Markets for recycled materials are notoriously volatile, and it often costs more to recycle waste than it does to simply bury it in a landfill. Recycling supporters, however, say the benefits of recycling far outweigh its drawbacks and predict a strong market for recycled materials in the future.

©PhotoDisc

THE ISSUES

3
- Do the environmental benefits of recycling outweigh the costs?
- Do government recycling mandates impede the creation of efficient markets for recyclables?
- Is a "pay-as-you-throw" system a more rational approach to waste management than other programs?

BACKGROUND

11 **Birth of a Movement**
In 1962, Rachel Carson's best-seller *Silent Spring* warned of environmental problems caused by toxic materials dumped in landfills.

11 **Federal Role**
Although recycling is a local matter, several federal laws have aided efforts, including the 1965 Solid Waste Disposal Act.

12 **Landfill 'Crisis'**
In 1987, the saga of a garbage scow without a port raised unfounded fears that the country was running out of room to bury its garbage.

CURRENT SITUATION

13 **An Uncertain Market**
Recycling today is a $30 billion industry, but market conditions and prices fluctuate widely among different recyclables.

16 **Communities Pitch In**
State and local budget cuts are prompting many communities to reassess their recycling programs.

OUTLOOK

17 **No End of Trash**
The amount of trash is expected to continue growing, and the recycling rate is expected to reach 35 percent of the waste stream by 2020.

SIDEBARS & GRAPHICS

4 **Recycling Efforts Span Wide Range**
Twelve states recycle at least 30 percent of their municipal solid waste.

6 **Our Throwaway Society**
Americans discard enough aluminum every three months to build the nation's air fleet.

8 **Paper and Lawn Trimmings Comprise Most of Trash**
Corrugated boxes make up the biggest category.

10 **Ups and Downs of Newspaper Recycling**
Prices fluctuate widely.

14 **How Germany Copes With Success**
It places the onus on industry.

19 **Chronology**
Key events since the 1940s.

20 **At Issue**
Does recycling make economic sense?

FOR FURTHER RESEARCH

21 **Bibliography**
Selected sources used.

22 **The Next Step**
Additional articles from current periodicals.

Note: For more information on this topic, please see the following pages in Press and Siever's *Understanding Earth,* Third Edition: pp. 272–274, 508; and in Merritts, De Wet, and Menking's *Environmental Geology:* pp. 17–19, 132–133, 168, 250–256.

The Economics of Recycling

By Mary H. Cooper

The Issues

Reduce, reuse, recycle. Since the first Earth Day in 1970, the mantra of the environmental movement has prescribed a simple remedy for the country's growing mountain of waste: If Americans would simply reduce the volume of stuff they buy, reuse what they have and recycle the rest, the depletion of natural resources would be slowed and there would be fewer potentially toxic garbage dumps.

Nearly three decades into the war on waste, however, most Americans have proved reluctant soldiers. There is little evidence that we are reducing our consumption of goods. The U.S. economy is in its eighth year of uninterrupted growth, driven largely by domestic consumption. If Americans were heeding the call to reuse what they have, environmentalists argue, they wouldn't be buying so many new products or tossing so much out. Although the sheer volume of trash is growing more slowly today than it did in the past, the United States remains the world's leading throwaway society. [1]

"Our per capita waste is just so out of whack," says Michele Raymond, whose firm, Raymond Communications Inc., tracks state recycling efforts. "We're just 20 percent of the people in the world, but we consume 80 percent of the world's resources. Our per-capita waste is the highest in the world and about twice the level of Germany and the United Kingdom. We're just trashing more than anyone else."

The only part of the anti-waste message that has taken hold to any noticeable degree is the call to recycle. Since 1988, when the Environmental Protection Agency (EPA) first set a recycling goal for the United States at 25 percent of total waste, communities across the country have introduced more than 8,000 curbside recycling programs and more than 3,000 composting facilities, all aimed at reducing the amount of household trash that ends up in landfills and incinerators. Colorful recycling bins are now a familiar sight on neighborhood streets.

"When I announced the 25 percent goal, only about 12 percent of the nation's garbage was recycled," says former EPA Assistant Administrator J. Winston Porter, now president of the Waste Policy Center, a research and consulting firm in Leesburg, Va. "The recycling rate grew pretty rapidly, and our goal was reached in 1995." The nation's overall recycling rate today is 27 percent of the total municipal solid waste stream. Sixteen percent is burned, while the remaining 57 percent ends up in landfills. [2]

Communities have been recycling trash for decades. Most early programs used drop-off points where people could leave their used newspapers, glass bottles and metal cans for a recycler to pick up every week or so. More than 9,000 drop-off centers are still in operation. But lackluster participation in drop-off programs prompted many communities to begin adopting curbside programs in the late 1980s as a way to boost recycling rates. For several years, the supply of recyclable materials was adequate for the infant recycling industry. Indeed, newspaper publishers created such a demand for old newsprint that "garbage rustlers" prowled neighborhood streets to collect junked papers from curbside bins.

But markets for recycled materials, much like those for pork

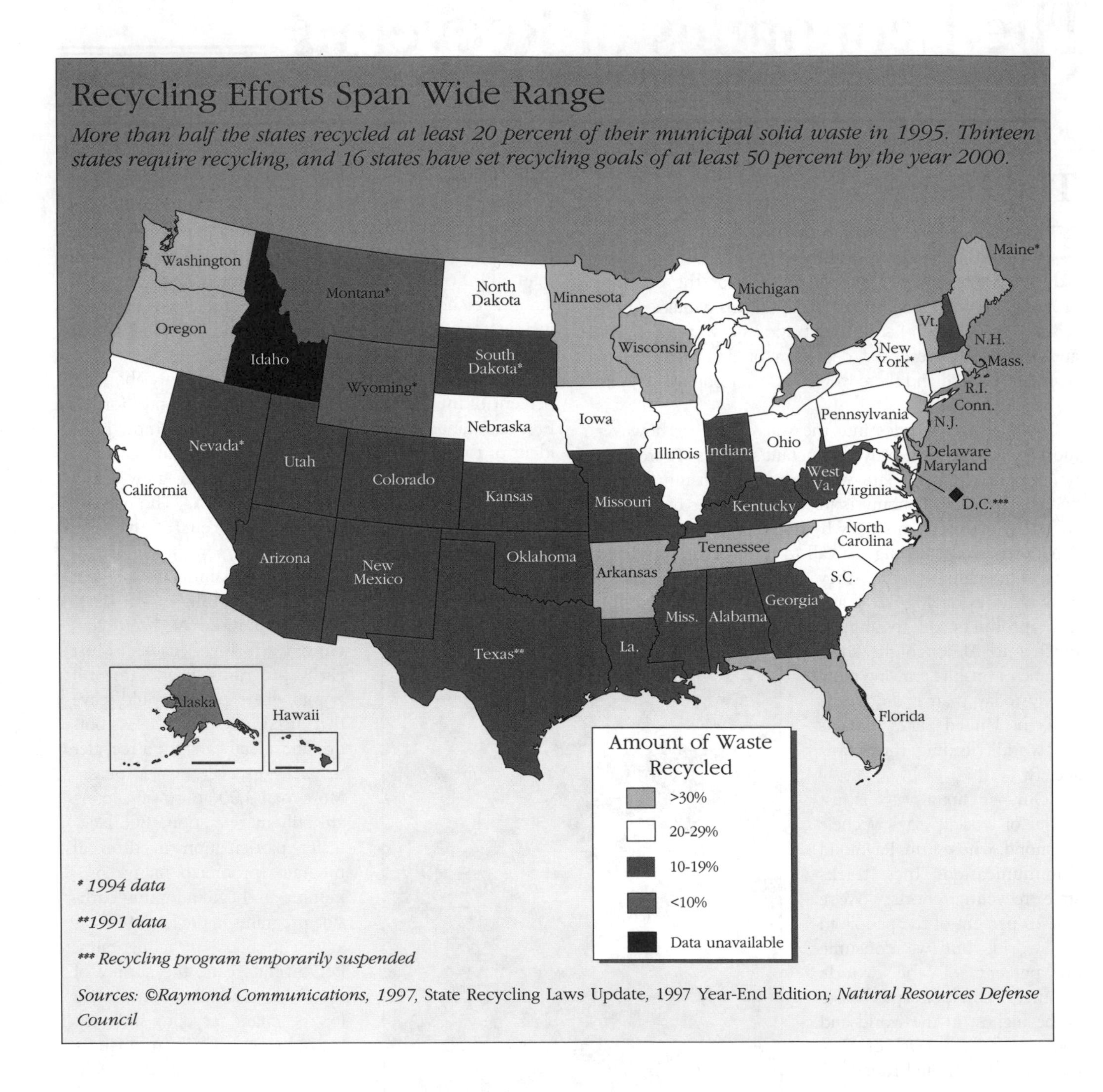

Recycling Efforts Span Wide Range

More than half the states recycled at least 20 percent of their municipal solid waste in 1995. Thirteen states require recycling, and 16 states have set recycling goals of at least 50 percent by the year 2000.

** 1994 data*

***1991 data*

**** Recycling program temporarily suspended*

Sources: ©Raymond Communications, 1997, State Recycling Laws Update, 1997 Year-End Edition; *Natural Resources Defense Council*

bellies, corn and other commodities, are notoriously volatile. Prices often gyrate when unusual weather, technological advances or other events cause sudden gluts or scarcities of a given commodity. When an oversupply of paper caused the booming market for recycled paper to collapse in the mid-1990s, a number of paper-processing facilities went under, and some recycling operators simply delivered the papers to the local landfill. Since then, recycling has come under closer scrutiny.

Critics blamed the collapse on state and local government goals and mandates for recycling. "Recycling may be the most wasteful activity in modern America: a waste of time and money, a waste of human and natural resources," wrote journalist John Tierney in a scathing criticism of recycling that appeared in *The New York Times Magazine* two years ago.[3] He and other skeptics charge that recycling programs fail to appreciably help the environment while imposing an unwarranted government intrusion into people's lives and disrupting the economy.

Tierney's criticism prompted a flurry of angry responses from readers and experts alike. "If we are serious about lowering the costs of recycling, the best approach is to study carefully how different communities improve efficiency and increase participation rates—not to engage in debating-club arguments with little relevance to the real-world problems these communities face," write Richard Denison and John Ruston of the Environmental Defense Fund (EDF). "By boosting the efficiency of municipal recycling, establishing clear price incentives where we can and capitalizing on the full range of environmental and industrial benefits of recycling, we can bring recycling much closer to its full potential."[4]

"If you ask people if they recycle, you're basically asking them if they care about the Earth, or if they like animals. They're going to say 'sure' because they don't want to sound like cretins."

Jerry Taylor—Director, natural resource studies, Cato Institute

Polls indicate that public opinion favors the environmentalists when it comes to recycling. Sixty-three percent of respondents to a recent survey reported that they were "personally doing more now" to help the environment, mainly by recycling.[5]

Skeptics dismiss such statistics, however. "I don't buy them for a second," says Jerry Taylor, director of natural resource studies at the Cato Institute, a Washington think tank that promotes free-market policies. "If you ask people if they recycle, you're basically asking them if they care about the Earth, or if they like animals. They're going to say 'sure' because they don't want to sound like cretins." Whatever the reason, however, the polls suggest a high degree of public interest in recycling.

Many experts in solid waste management say the debate over recycling has been cast in overly simplistic terms that fail to accommodate the complexities of the markets for different recyclable materials. Whether it makes economic or even environmental sense to recycle, they say, depends on the material in question. "Recycling is like a vacuum cleaner that sucks up dispersed materials and reuses them," says Lynn Scarlett, vice president for research at the Reason Foundation, a nonprofit think tank in Los Angeles, Calif. "Where the material is of uniform quality, collected in large quantities and easy to isolate from contaminants, there are net benefits to recycling."

Steel scrap is one such recycling success story. Used cars, appliances and other goods that end up in junkyards are broken apart to allow giant magnets to separate the steel from plastics and other contaminants; then crushers collapse and bale the metal. Electric arc mini-mills melt the metal down to produce high-quality recycled steel, at less cost than "virgin" steel and ready to be made into new products. Steel has been recycled profitably for years, without the benefit of government support, and now accounts for about half of all steel consumed, Scarlett says.

The economic picture for materials collected in curbside programs is mixed. "Tin" cans, which actually are largely steel, are profitably recycled like steel scrap. Recycled aluminum beverage cans also are a hot commodity because they can be processed into new cans at less cost—and with less pollution—than mining and processing bauxite into virgin sheet aluminum. And despite the market's volatility, there usually is strong demand for old newspapers and cardboard boxes.

But the case for recycled plastics is more complex. There are six major types of plastic resins found in consumer products, only two of which are relatively easy to recycle. Containers made with these resins often have to be manually separated out of the recycling bin and carefully washed to remove contaminants before they can be reprocessed, an expensive process that is not always cost-effective.

"Environmentalists are right to say there are many opportunities for recycling," Scarlett says. "But for some products, it makes little sense to have any recycled content. Like any manufacturing process, recycling is a way of making products, an alternative to using virgin feed stocks. The question is whether you can do it and get the product you want at a cost you want. In many cases the answer is yes, in others no. Each product has its own story to tell."

Nonetheless, a growing number of Americans have incorporated recycling into their daily lives and expect curbside service as a basic amenity, like electricity and other utilities. Supporters predict that recycling will keep expanding if Americans are made more aware of the benefits of waste reduction. To that end, more than 1,000 communities participated in the first America Recycles Day last Nov. 15. Sponsored by the EPA and environmental organizations, the event was designed to encourage recycling and the use of products containing recycled materials.

"For many people across the United States, recycling is a matter of habit, something they do in the course of their daily lives, both in the office and at home," says Richard Keller, chief of recycling at the Maryland Environmental Service, a state agency that sorts and markets recycled materials. "But recycling is at a fairly critical crossroads right now. We're already recycling the materials that are easy to recycle. The next step is going to depend on whether we start dealing with materials that are hard to recycle." These include more grades of plastics, as well as organic materials such as food wastes, paper food wrapping, diapers and tissues, which could go to special composting facilities.

The success of recycling programs varies widely according to local conditions. In some highly populated regions of the Northeast and the West, high fees for landfill dumping make recycling especially attractive. In rural areas of the Rocky Mountain West, the high cost of collecting materials and the low cost of land for dumping waste have hindered recycling efforts. Ultimately, it is up to consumers and their elected officials in state and local governments to decide how best to dispose of their trash. These are some of the issues that shape their decisions:

Do the environmental benefits of recycling outweigh the costs?

Some markets for recyclable materials are strong enough to pay for their collection and conversion into new products. Recycled steel from cars and appliances is one of the most profitable post-consumer materials. Recovery of steel and aluminum cans also tends to more than pay for itself.

But demand for other products is less predictable. Plastics are hard to recycle economically because there are so many types of materials, and most are costly to clean and

Our Throwaway Society

- *Every week more than 500,000 trees are used to produce the two-thirds of newspapers that are never recycled.*
- *Americans throw away enough office and writing paper annually to build a wall 12 feet high from Los Angeles to New York City.*
- *Every year Americans dispose of 24 million tons of leaves and grass clippings, which could be composted to conserve landfill space.*
- *Americans throw away enough glass bottles and jars to fill the 1,350-foot twin towers of New York's World Trade Center every two weeks.*
- *American consumers and industry throw away enough aluminum to rebuild our entire commercial air fleet every three months.*
- *Americans throw away enough iron and steel to continuously supply all the nation's automakers.*
- *Americans use up 2.5 million plastic bottles every hour, only a small percentage of which are now recycled.*

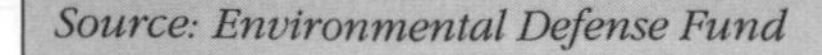

Source: Environmental Defense Fund

return to a usable form for new consumer products. Paper is less difficult to recycle, but the market for recycled paper has fluctuated wildly in recent years. After peaking in 1995, prices of recycled newsprint sank so low that some recycling contractors simply dumped the paper they collected in landfills.

On average, curbside recycling programs tend to cost slightly more than they earn from the sale of collected materials. According to Franklin Associates Inc., a Prairie Village, Kan., research firm that conducts solid waste studies for the EPA, residential recycling programs cost on average $2 a month per household. "The cost varies widely from community to community," says Bill Franklin, the firm's chairman. "But recycling costs are a very small percentage of the total cost of solid waste removal, which averages $10 a month per household." Commercial recycling, which generally is contracted out by businesses to private haulers, probably makes more money than it costs. "We don't have good numbers on commercial recycling," Franklin says. "But it must be cost-effective, or they wouldn't bother to do it."

Environmentalists say the quantifiable financial costs of collecting, sorting and processing recyclable materials are far outweighed by an array of benefits to the environment. Nobody disputes the fact that recycling reduces the amount of trash that ends up in landfills or incinerators. Of the 208 million tons of municipal solid waste generated in the U.S. in 1995, 27 percent, or more than 56 million tons, was recycled. Fifty-seven percent ended up in landfills, and the remaining 16 percent was burned. [6]

But some critics dispute the importance of saving landfill space. "It's not in the least bit true that we're running out of places to put our garbage," Taylor says. He cites a Cato Institute estimate that all the trash generated in the United States over the next 1,000 years would fit into a single, 30-square-mile landfill 1,000 feet deep. "Of course, nobody is going to build that big a landfill," Taylor says. "But this shows the idea that we're running out of places to put garbage is just silly. Anyway, most landfills, when they're retired, are sodded over and turned into golf courses or other public facilities."

Many experts say recycling does far more to help protect the environment than merely preserve land that would otherwise be needed for waste disposal. "The more important goals of recycling are to reduce environmental damage from activities such as strip mining and clear-cutting and to conserve energy, reduce pollution and minimize solid waste in manufacturing new products," write Denison and Ruston of the EDF. "[R]ecycling is an environmentally beneficial alternative to the extraction and manufacture of virgin materials, not just an alternative to landfills." [7]

©PhotoDisc

Lackluster participation in drop-off programs prompted many communities to adopt curbside collection programs to boost recycling rates.

But some experts say the environmental benefits of recycling depend on the material in question. "Each material has its own tale to tell," Scarlett says. "It takes about 95 percent less energy to make a new can out of recycled aluminum than to make can sheeting by mining bauxite and smelting the ore. But the energy savings offered by recycled glass are very modest because it takes only slightly less heat to process cullet [recycled glass] than to make glass from silica. And if you have to transport the glass hundreds of miles to a reprocessing facility, those gains may be undone because glass is very heavy and consumes a lot of energy to transport."

Even aluminum recycling can harm the environment under some circumstances. "If you have to drive trucks far out into the countryside to pick up the cans, the air pollution generated by the trucks will outweigh the pollution saved by

Garbage Loaded With Paper, Lawn Trimmings

Cardboard boxes, newspapers and yard trimmings comprised nearly two-thirds of the nation's municipal solid waste recovered in 1995. Some recycled materials, such as batteries and aluminum cans, had high recovery rates but contributed only a small percentage of the total materials recovered.

Recovery of Products in Municipal Solid Waste, 1995

Product	Tons Recycled (in thousands)	Percent of Product Recycled	Percent of All Solid Waste Recovered
Corrugated boxes	18,480	64%	33%
Yard trimmings	9,000	30	16
Newspapers	6,960	53	12
Glass bottles and jars	3,140	27	6
High-grade office papers	3,010	44	5
Major appliances	2,070	61	4
Lead-acid batteries	1,830	96	3
Steel packaging	1,550	54	3
Aluminum beverage cans	990	63	2
Magazines	670	28	1
PET soft drink bottles	300	46	1
HDPE milk and water bottles	190	30	<1
All other products	8,000	8	14
Total recovery	**56,190**	**27%**	**100%**

Source: Environmental Protection Agency, "Characterization of Municipal Solid Waste in the United States," 1996 Update

recycling," Porter says. "If you try to get the last squeal of the pig, you end up doing more environmental damage than good. While the environmental benefits of recycling outweigh the costs for most things, people who treat recycling as a religion haven't considered the whole picture." [8]

Some critics go further, saying the environmental benefits of recycling have been grossly exaggerated. "You'd be hard-pressed to find real environmental benefits in recycling under any circumstances," Taylor says. "If you're recycling glass, what commodity are you saving, sand? We're not running out of sand. We're also not running out of energy—energy prices are the lowest ever, adjusted for inflation. And while, as a general matter, you can argue that energy consumption is a precursor of industrial pollution, if we suddenly started recycling everything instead of using virgin materials, the reduction in energy use wouldn't be all that dramatic."

Do government recycling mandates impede the creation of efficient markets for recyclables?

Most curbside recycling programs now in effect were introduced in the wake of the "landfill crisis" of the late 1980s, when several cities in the Northeast appeared to be running out of space to bury their garbage. Although the perceived crisis never materialized, governments at all levels called for increased recycling. In 1988, when the nationwide recycling rate stood at about 12 percent, then-EPA Assistant Administrator Porter called for Americans to boost that rate to 25 percent by 1993. States followed suit by setting goals of their own, prompting local governments to create or expand curbside residential recycling programs.

The majority of recycling mandates are merely goals, with no enforcement provisions. "Most recycling in this country is voluntary," Raymond says. "Some states have mandatory curbside separation, and some have mandatory goals, but most of it is voluntary."

Some jurisdictions, however, including California and New York City, impose fines for non-compliance with recycling mandates. In addition to recycling goals, some governments require agencies to buy a specified percentage of recycled goods, such as office paper, plastic traffic cones or paving materials made of old tires. At the federal level, for example, President Clinton issued an executive order in 1993 requiring all federal agencies to buy printing and writing paper with at least 30 percent recycled content by the end of 1998.

There is little doubt that government mandates have spurred growth in recycling programs. About 40 states have set recycling goals, ranging from 15–70 percent of the waste stream, according to Keller. "Most programs call for 25–50 percent recycling over varying periods of time," he says. "I think that without those mandates, especially on the residential side, you'd see far lower rates of recycling than what you're seeing now."

But critics say mandates distort the markets for recycled materials, possibly impeding their development over time. "Post-consumer material is a resource just like water, energy or any other resource, and a certain amount of it is going to be reused because it makes economic sense to reuse it," Taylor says. "But there's a lot of it that it doesn't make economic sense to reuse, and that's where government mandates come in. It cannot make economic sense to mandate the use of a material that nobody would otherwise use. You can build a nice Potemkin village marketplace out of that, but it doesn't mean you're really helping the economy, because the money used to pay for recycling programs is money that would otherwise have gone to more productive uses."

Many critics blame recycling mandates for the collapse in recycled paper prices in 1995. By suddenly increasing the supply of recycled paper in the early 1990s, this argument goes, governments flooded the market with more material than reprocessors could absorb. But Keller, who finds buyers for his agency's recycled materials, says mandates do not significantly distort the market, at least over the longer term. "In Maryland, which has set recycling goals of 20 percent for large subdivisions and 15 percent for small ones, the marketplace is very resilient," he says. "There may have been some marketplace dislocations when the programs came on line, but as a matter of fact there are some industries that we're working with that are scrambling to find materials. You have to recognize that the markets fluctuate."

Most curbside programs collect recyclables from individual houses, making them most common in urban and suburban neighborhoods as well as residential areas in smaller communities. Typically, apartment houses and commercial businesses are not included in public recycling programs. Some experts charge that by focusing on households, which generate relatively small quantities of trash, governments targeted the least suitable population for efficient recycling programs. "If they had actually studied where the trash was coming from, they would have put the mandates on industrial plants first, then commercial businesses and then multifamily housing units," Raymond says. "Curbside recycling would have come last in order of priority."

Like many other experts, Raymond says mandates have a positive role to play in helping markets for recyclables get started. But in her view, the focus on curbside collection has actually stymied that effort. "Had they phased curbside collection in last, these markets might

"Recycling programs don't encourage you to buy less, use both sides of the paper for photocopies or do other things to reduce the amount of stuff consumed. Variable rates does that—it adds something that even mandatory recycling can't accomplish."

Lisa A. Skumatz, president, Skumatz Economic Research Associates

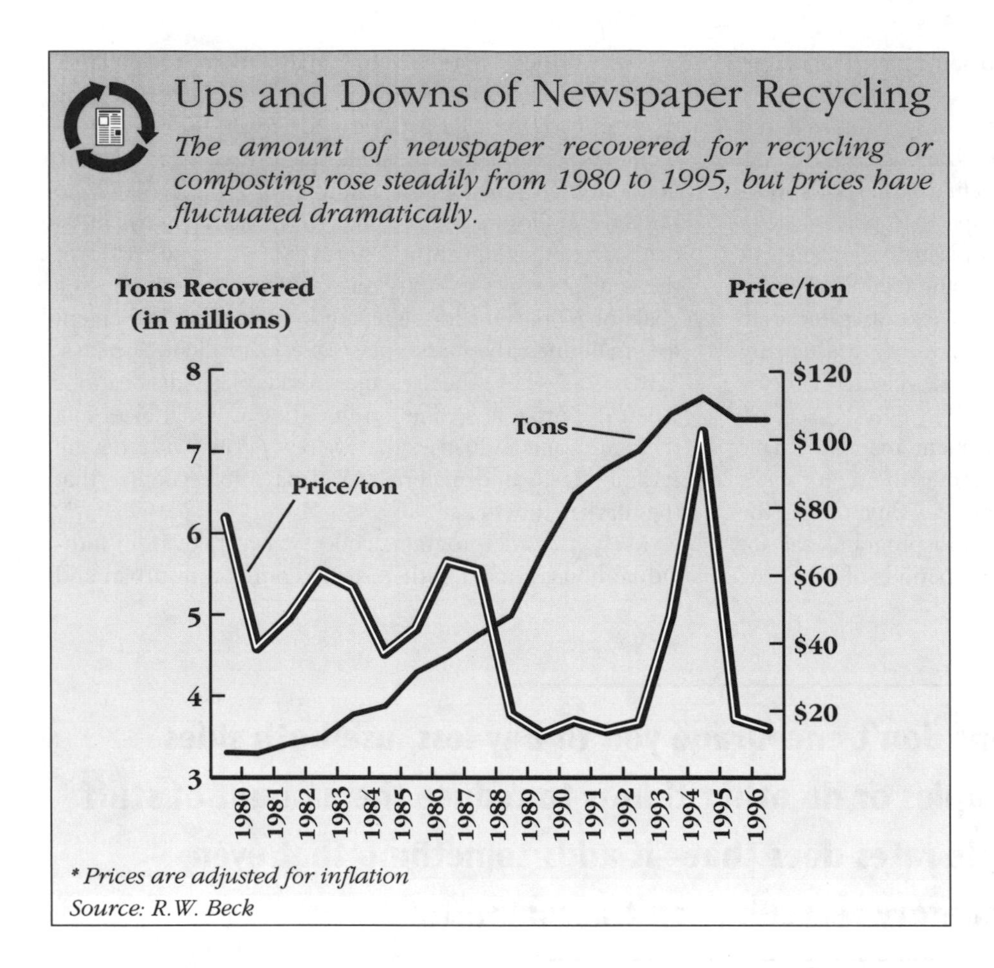

Ups and Downs of Newspaper Recycling

The amount of newspaper recovered for recycling or composting rose steadily from 1980 to 1995, but prices have fluctuated dramatically.

** Prices are adjusted for inflation*
Source: R.W. Beck

have stabilized, and they would know what makes sense to pick up and what doesn't," Raymond says. "With commercial recycling, if the market goes bad for paper they simply stop recycling. But you can't do that with curbside recycling because it takes so long to educate people about what stuff to put in their bins. You can't just drop the glass and continue collecting the plastics, even though that probably would be better for the environment in the long run. With residential recycling, once you turn on that spigot you can't turn it off again."

Is a "pay-as-you-throw" system a more rational approach to waste management than other programs?

Experts agree that it is virtually impossible to educate consumers to periodically adjust the list of materials they leave for curbside pickup in response to fluctuations in the markets for recyclables. But program operators have found other ways to make recycling more cost-effective. Costs can be trimmed by shifting pickup schedules from weekly to every other week, investing in automated trucks to reduce labor costs or simply redesigning collection routes to use fewer trucks or reduce the time needed to complete collection.

But the most promising cost-saving measure may be variable-rate waste management. Also known as pay-as-you-throw, this approach replaces the flat fee typically charged households for waste removal, including recyclables, yard trimmings and other trash, with fees that vary according to the amount of trash destined for the landfill or incinerator. "Pay-as-you-throw is an excellent program that bears consideration in most communities," says Lisa A. Skumatz, president of Skumatz Economic Research Associates in Seattle, who found the system raised recycling rates by up to 10 percent in more than 500 communities she studied. [9] "It can be extremely cost-effective, the biggest-bang item you can get to improve recycling and yard-waste diversion." She estimates that more than 5,000 communities have adopted variable-rate programs to date.

Pay-as-you-throw works by charging households for the waste they generate, and so provides an economic incentive to recycle or compost as much trash as possible. Customers usually are charged by the garbage can or by the bag for non-recyclable trash. In Seattle, for example, customers pay by the container. The program is not without glitches: Customers often jam as much as they can into one container—a practice so common it's known as the "Seattle stomp." "Some guy actually hurt his back jamming stuff into his trash barrel and sued the city," Franklin says. "But in general, the pay-as-you-throw system does seem to work. If we know there's a penalty for the second barrel of trash, the economic incentive helps us manage waste better."

Some communities have made the system more accurate by using special trucks equipped with scales to weigh the trash they collect. "The variable-rates approach works for the customer because it says you can do what you want," Skumatz says, "but you'll save real dollars if you recycle."

By placing greater responsibility on the customer, variable rates not only encourage people to recycle but also to reduce their consumption. "Recycling programs, whether they're mandatory or not, in no way encourage source reduction in the first place," Skumatz says. "They don't encourage you to buy less, use both sides of the paper for photocopies or do other things to reduce the amount of stuff consumed. Variable rates does that—it adds something that even mandatory recycling can't accomplish."

But there is a downside to variable rate-systems. "While unit-based pricing rewards people who are more efficient, it's not a total panacea," Porter says. "People will haul their garbage to the Dumpster in the McDonald's parking lot on their way to work, or they'll sort their mail in the post office and throw away the junk mail before going home. So there's a lot of trash shifting going on."

Porter also questions the ability of most people to drastically reduce the amount of stuff they consume. "What exactly are you supposed to do, stop buying milk or feeding the dog?" he asks. "You may get a 10 percent reduction in consumption with pay-as-you-throw, but not much more. People can't change their habits that much."

Variable-rate programs also have limited impact beyond the supply side of the market for recyclables. "Pay-as-you-throw is the best idea we have now because it provides the correct signal to the consumer for wasting less," Raymond says. "The problem is, it doesn't solve the problems we have with recycling right now because it doesn't create markets. It just changes consumers' habits and improves their cooperation in putting all of their recyclables out. Now what are you going to do with them? You've got to have markets for them. Industry has to use recycled materials for the markets to operate properly."

Other critics object to pay-as-you-throw for the same reasons they object to all recycling programs. "I'm suspicious whenever governments at any level think they know exactly the right way to do things," Taylor says. "Some people will be annoyed at having to pay for each bag they throw out and would far rather pay a flat fee for trash removal. I don't think government should be involved in recycling at all. Each household should be allowed to make its own decisions about who's going to collect their garbage and under what terms."

Background

Birth of a Movement

Throughout history, recycling has been the rule rather than the exception. Even after the Industrial Revolution ushered in the modern era of mass production and consumption of consumer products, many basic materials continued to be recycled. Until the late 19th century, for example, paper was made from rags, and the demand for rags upheld a robust market for used clothing and other textile scraps. [10] Scavengers picked garbage dumps for these and other materials that were routinely fed back into the production cycle. Today, the remnants of older types of recycling persist in the form of scrap dealerships where cars, appliances and other goods are broken up, sorted into their steel and other metal components and sold to mini-mills for use in new products.

Although recycling had always been a common business activity, it was not until the 1940s that people began collecting materials for reasons other than profit. During World War II, Americans voluntarily contributed metals, rubber and other materials for the war effort. More than a third of all paper and paperboard products, as well as other materials, were recovered and recycled by the war industries.

Recycling fell from view after the war, but re-emerged in the 1960s in response to a new concern—the environment. Lady Bird Johnson, the wife of President Lyndon B. Johnson, helped launch a campaign to clean up the trash strewn along the nation's roadways, and Rachel Carson, in her 1962 best-seller *Silent Spring*, warned of the dangers of toxic chemicals left in landfills. Largely unregulated at the time, landfills were frequently used as dumps for hazardous wastes, which leached into the groundwater. [11] Recycling was seen as a way to reduce the amount of trash and thus the need for more landfills.

By the early 1970s, thousands of grass-roots recycling centers had appeared where consumers could drop off newspapers, glass bottles and aluminum cans. In addition, by 1974 more than 100 communities had set up curbside collection services for recyclables, mostly newspapers. Newspaper collection fell off following a slump in the newsprint market in the mid-1970s. But the demand for recycled aluminum remained strong; by 1975, a quarter of all aluminum cans were being recycled.

Federal Role

Although recycling has always been a matter of state and local jurisdiction, the federal government has enacted several laws since the environmental movement's inception in the 1960s that have indirectly influenced the progress of recycling programs. [12] The 1965 Solid Waste Disposal Act launched federal research into the technology of waste disposal and provided technical and financial assistance for state and local waste programs. The 1970 Resource Recovery Act extended the federal effort to include recycling programs and waste incineration, including waste-to-energy facilities, which generate steam or electricity from burning garbage. [13] The law stopped short, however, of

Of the 208 million tons of municipal solid waste generated in the United States in 1995, 57 percent ended up in landfills.

©PhotoDisc

setting standards, and it required states to come up with plans for managing their waste as a condition for receiving the federal aid.

The 1976 Resource Conservation and Recovery Act (RCRA) set national standards for landfills and incinerators that were strengthened in 1984. While the law did not address recycling directly, its requirements of environmental safeguards, such as landfill liners and smokestack filters, made it more expensive to bury or burn waste and thus encouraged recycling indirectly.

Waste incineration gained support as an attractive source of energy in the wake of the energy crises of the 1970s. In 1978, Congress encouraged the development of waste-to-energy facilities with the Public Utilities Regulatory Policies Act (PURPA), which required utilities to buy power from these plants at favorable rates. Enthusiasm for waste-to-energy facilities soon cooled, however, out of concern over the toxic ash they generate, which must be disposed of in special landfills. As a result, only about 16 percent of municipal solid waste is incinerated. [14]

Landfill 'Crisis'

The administration of Ronald Reagan (1981–1989) enacted no major laws dealing with solid waste. But public perceptions that the country's mounting trash flow posed threats to the environment continued. These fears escalated in 1987, when the *Mobro 4000*, a garbage scow, sailed for days down the East Coast and into the Caribbean Sea in search of a port to dump its fetid cargo. The widely publicized voyage came to symbolize the emergence of a landfill "crisis," in which cities, having run out of space to bury their own garbage, were desperately trying to export it to other jurisdictions.

As it turned out, the *Mobro's* problems had little to do with the lack of landfill space. The barge operator had simply set sail without signing any agreements with landfill operators to dump its garbage, and port authorities turned the scow away, not because local landfills were full but because they feared it carried hazardous wastes. [15] But the *Mobro's* wayward journey bolstered public support for recycling as a solution to the problem.

By this time, recycling operations had expanded in number and in the range of materials collected. Beginning in the early 1980s, recycling operators built "material-recovery facilities" where they could sort, bale and market the recyclable materials they collected in curbside programs. In the absence of federal standards for recycling, states set their own standards. Today, most states have passed laws aimed at reducing the total waste stream and encouraging recycling as well as composting of yard trimmings. [16]

Current Situation

An Uncertain Market

Community recycling programs are just the first link in the recycling process. The next role is played by markets for recyclable materials—the paper mills, steel mills, aluminum smelters, glass container makers and plastics reprocessors that transform these materials into new products.

The 1990s have seen both unprecedented growth in recycling programs and volatility in the markets for recyclable materials. A glut in the supply of recyclables that followed the sudden increase in curbside programs in the late 1980s caused a crash in prices in 1991, as new manufacturing plants failed to keep pace with the supply of raw materials. A so-called "basket" of recyclables that fetched $70 a ton in 1989 went for just $25 a ton by 1992, prompting some communities to pay haulers to cart away the materials they had collected. [17]

Barely two years later, however, prices soared. Old corrugated cardboard rose in value from $25 to $150 a ton during the first six months of 1994, while mixed paper prices skyrocketed from $5 to more than $200 a ton. Prices for old newsprint, plastic soda bottles and milk jugs and aluminum cans also soared. [18] Boom again turned to bust in mid-1995, when prices for paper and plastics plummeted.

Such extreme swings in the market come as no surprise to recycling experts. "After all, these are market commodities, and how they fare depends on what is going on in the overall economy," Keller says. "If the economy goes south and people stop making as many houses and automobiles and buy less retail goods, you'll see a drop-off in the end market for recyclables."

©PhotoDisc

After peaking in 1995, prices for recycled newsprint sank so low that some contractors simply dumped the paper they collected in landfills.

Today, recycling in the United States is a $30 billion industry. As the industry has grown, the market for recyclables has become somewhat more stable, and while prices for most materials are nowhere near their peaks of three years ago, they are close to historical norms. According to R.W. Beck Inc., a consulting and engineering firm that tracks the market in recyclables, the value of a basket of nine materials it measures stood at an estimated $80 a ton in 1997, much lower than the $145 reported in 1995 but higher than the $68 reported in 1996. [19] A closer look shows how different market conditions are for the most commonly recycled materials:

Paper—Almost 40 million tons of paper were recovered from the waste stream in 1996, or nearly 85 percent of all recycled materials. More than half of that was corrugated containers, the single largest item in Beck's index. These are primarily boxes used for packaging and shipping goods from factories to retailers, and most are collected through privately contracted recycling programs. Demand for this material is generally strong.

The glut of old newspapers that sent the market into a tailspin three years ago has abated, in part because growth in newspaper circulation has slowed with the advent and growth of electronic news media. Mixed paper, one of the more recent items included in curbside collection, is expected to be in greater demand as an alternative to corrugated boxes for the production of paperboard, used to make shoeboxes, cereal boxes and other items. Likewise, high-grade paper is in demand, in part because of federal and state purchasing mandates requiring public agencies to buy writing and printing paper with recycled content.

Prices for recycled paper in 1997 ranged between $5.69 per ton for mixed paper to $109 per ton for high-grade paper. With the exception of old newspapers, which fell

How Germany Copes With Recycling Success

The United States may have launched the recycling movement, but other countries have embraced stricter recycling requirements, notably Germany. German manufacturers must take responsibility for collecting and recycling or reusing the packaging materials for the products they sell. Unlike U.S. recycling programs, which focus on consumers and for the most part rely on their cooperation, the German system embodies the "polluter-pays" principle. It holds that because industry is responsible for choosing what packaging it uses, it should bear the burden of recycling it.

The law was partly a consequence of the collapse of European communism. Before German reunification in 1990, West Germany had "solved" the problem of dwindling landfill space by exporting its trash to East Germany, where landfill regulations were minimal.[1] Meanwhile, West Germany's increasingly influential Green Party helped pass strong environmental protection regulations. With reunification, those regulations were extended to the whole country, and a new solution to Germany's waste problems had to be found.

The Ordinance on Avoidance of Packaging Waste, which took effect in June 1991, required industry to switch to reusable shipping cartons and reduce its use of non-essential "secondary" packaging, such as cardboard boxes containing bottles of aspirin or the plastic "blister" packs used to display many consumer goods. Beverage producers were able to comply with the law by using refillable bottles, a longstanding practice in much of Europe.

But problems quickly arose for essential, "primary" packaging materials, the countless pill bottles, toothpaste tubes and myriad other containers that manufacturers were required to take back and recycle or reuse under the law. Faced with a near-impossible obligation, German industries banded together and created a consortium—the Duales System Deutschland (DSD)—to provide collection and recycling services on their behalf. Manufacturers pay DSD for the right to place a green dot on their packaging, which exempts those items from the law's take-back provision. DSD then pays waste companies to collect those packaging materials from curbside bins or drop-off centers and sells the materials to industry for recycling.

The German system immediately boosted recycling rates. By 1994, DSD had grown into a multibillion-dollar business and was recycling two-thirds of the country's packaging materials. But the system has also been the victim of its own success, especially in the collection and recycling of plastics. Overwhelmed by a glut of collected plastic packaging, DSD exported much of the excess to other countries where it was often buried or burned.

Despite its drawbacks, the German recycling system is being emulated in other countries, particularly in Europe, where high population densities make landfills less feasible than in the United States. As part of an effort to harmonize regulations throughout the 15-member European Union, the EU in 1994 adopted a Packaging Directive setting both minimum and maximum recycling goals—the latter aimed at avoiding some of the German system's problems. By 2001, all member countries are to recycle from 25–45 percent of their packaging. Germany, which already exceeds the maximum quota for all materials, must either reduce its recycling rate or demonstrate that it can recycle its own materials.

A number of countries have adopted or are considering recycling programs based on the German green dot system. For example, Austria, Belgium and France have introduced similar programs, modified to allow incineration of recyclables as a way to avoid the oversupply of materials that has hampered the German program.

The German model may have little appeal, however, in the United States, where recycling has always been a largely voluntary effort based on individual participation. Additionally, the German approach owes much of its success to the country's high population density and limited geographical area. In the United States, with its vast distances and large tracts of undeveloped land, landfilling is still an option, while transportation costs discourage recycling in many parts of the country.

"Germany's packaging law is the most stringent in the world, and they spend up to $4 billion a year enforcing it," says Michele Raymond, publisher of *State Recycling Laws Update*. "It's a totally different system there."

[1] Information in this section is based on Frank Ackerman, *Why Do We Recycle?* (1997), pp. 108–109.

slightly in 1997 to $15 a ton, prices for recycled paper materials rose slightly from 1996 to 1997.

Although many deinking plants were built in response to the 1995 peak in paper prices, a number of paper recycling manufacturers have been in business for years. Marcal Paper Mills Inc., for example, has been using a variety of waste papers to make 100 percent recycled-content napkins, tissues and paper towels at its Elmwood Park, N.J., plant since 1947.

New deinking technology that can reprocess coated paper, such as old magazines and catalogs, has increased demand for a broader range of paper products than was once collected for recycling. But paper mills still report problems resulting from contamination of the materials, especially from "stickies," or adhesive notes, which raise the cost of producing recycled paper. [20]

Glass—After newspapers, glass containers are the most common item by weight found in residential waste. Although they are rapidly being replaced by plastics for soft drinks, glass bottles and jars are widely used to package beer, other drinks and food products. Recovery of glass containers has fallen slightly since it peaked in 1994. Prices for clear glass—the only type measured in Beck's analysis—fell 20 percent in 1997, to $37 a ton.

Clear, green and brown glass bottles have different properties and must be processed separately. In areas where they are collected together, breakage can quickly contaminate the recovered materials. Another obstacle to reprocessing glass is its weight, which makes it costly to ship, especially in sparsely populated regions such as the Mountain West. Transportation is becoming a greater factor in the cost of reprocessing glass because of consolidation: 127 regional plants have been replaced by 62 larger plants over the past 20 years. [21]

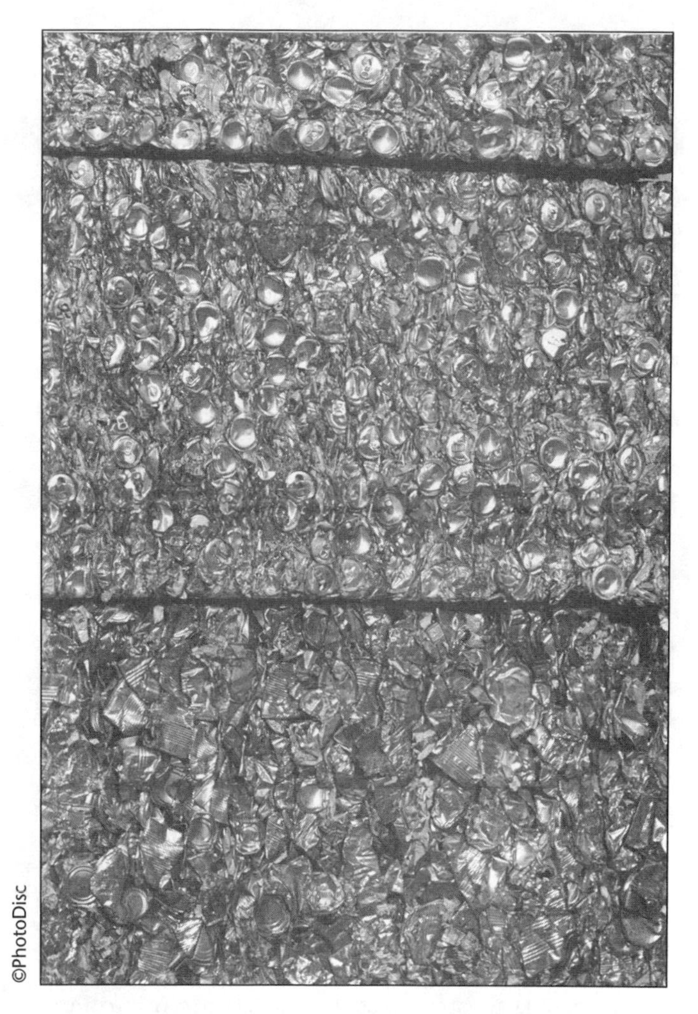

©PhotoDisc

Nearly two-thirds of all aluminum beverage cans were recycled in 1997. About a million tons of aluminum cans are recycled each year.

Steel cans—Scrapped appliances as well as steel cans collected curbside account for most steel and iron recovered from municipal solid waste. Almost 60 percent of all steel cans produced in the United States, mostly food cans, are recovered for recycling, and the percentage is growing steadily, thanks in part to the recent inclusion of aerosol cans in many curbside programs. Perhaps reflecting that trend, prices for baled steel cans fell to $50 a ton in 1997, the lowest level reported by Beck since 1980. However, given high demand for scrap steel by the steel industry in the United States and abroad and the recent increase in the number of electric-arc furnaces in operation, prices for recycled steel are expected to rise in coming months. [22]

Aluminum cans—At 64 percent, the 1997 recycling rate for aluminum beverage cans outstrips that for steel cans, and amounts to about a million tons a year. But the price of aluminum cans is much higher—$1,090 a ton in 1997—and has risen by more than half since 1993, making these the most valuable items collected in most curbside programs. Unlike most recyclables, aluminum is relatively unaltered by reprocessing, so it can be recycled over and over again with little sacrifice in quality. Aluminum cans are unusual also because they typically emerge from the recycling process as new aluminum cans, completing the closed loop depicted in the universal symbol designating products with recycled content. Domestic demand for aluminum cans is so high that less than 1 percent of them are exported to overseas processors.

Plastic—More than half of the nation's roughly 32,000 communities have access to programs that collect one or more types of plastic. [23] But plastic recycling has been hampered by the difficulty of separating the different types of containers and packaging materials and reprocessing them into materials suitable for remanufacture. Of the scores of plastic resins used in manufacturing, six are most commonly found in everyday consumer products. Of these, only two have been widely included in recycling programs—polyethylene terephthalate (PET), used to make beverage bottles, and high-density polyethylene (HDPE), used in milk jugs and other containers.*

Prices for old PET bottles, much like those for recycled paper, have fluctuated in recent years, ranging from a record high of $354 per ton in 1995 to $40 per ton in 1996. The main reason for the crash in prices for PET bottles was the opening of numerous factories here and abroad producing virgin resin, which curbed demand for scrap PET. Last year, Beck reports, PET prices rebounded to $118 per ton. There is no lack of capacity to reclaim PET—27 plants in 18 states processed more than 300,000 pounds of post-consumer PET in 1995. [24]

Both the recovery rate and prices for HDPE bottles have risen in recent years. Many recycling programs have expanded the list of eligible HDPE items to include water and juice bottles as well as pigmented bottles used for liquid detergents. Baled HDPE containers, which totaled 660 million tons, brought $421 a ton last year, almost as much as during the 1995 peak. Recycled HDPE is often used to make new containers, plastic plumbing pipes, leaf bags and plastic "lumber."

The complexity of plastics recycling makes it hard to develop stable markets for products made of resins other than PET and HDPE, however. New technologies are making it possible to break plastic materials down into their original state, so that they are virtually indistinguishable from virgin resins, but most resin producers have not invested in this technology. [25] "Remember, these producers are the Exxons of this world, large, traditional companies that are not likely to spend money for a small niche business like recycling," Raymond says. "Anyway, why should they want to use less virgin resin? They're selling more and more of it, 40 percent more over the past six years alone. Plastics surpassed steel in the 1970s in terms of volume. Today it's bigger than just about anything."

Most of the plastic that is recycled undergoes only partial reprocessing before it is transformed into new products, products that bear little resemblance to the bottles and packaging materials consumers tossed in the bin. In such an "open-loop" application, plastic bottles are washed, ground up and made into fiber for fleece jackets and carpets, or they are exported. Only about 16 percent is used to make new bottles. [26]

*Plastic containers are numbered from 1–7 depending on which resins they contain.

Another obstacle to further plastic recycling is the growing popularity of single-serve PET drink bottles. Because they are often used and discarded away from home, these bottles often elude recycling programs and end up in the trash. Businesses and public agencies are beginning to install special bins to encourage recycling of these bottles.

Compost—Though it is collected separately from other recycled items, yard trimmings are commonly picked up curbside and hauled to a municipal composting facility. Once composted, leaves, grass clippings, branches and some food wastes produce a nutrient-rich soil additive that is purchased by nurseries, landscapers and residential gardeners. But because its value is so low in most parts of the country, the cost of collecting yard wastes is usually borne by consumers and embedded in the bill for solid waste removal.

Some communities provide special bins to encourage households to compost their own organic wastes and keep this material out of landfills. In the Northeast and other areas where landfill operators charge high "tipping" fees to dump trash, yard wastes are actually banned from the waste stream.

Communities Pitch In

The 1990s have seen rapid growth in recycling efforts. Today, about 27 percent of the nation's municipal solid waste is being recovered through recycling, up from just 17 percent in 1990. But because materials that are relatively easy to recycle are already being collected by many programs, the pace of new program start-ups has slowed over the past year, a trend that experts predict will continue.

Recycling often pays for itself in peak years, such as 1995, when communities earn more from selling materials than they spend to collect them. But when the value of recyclables drops, as it did in 1996, recycling often poses a financial burden on local governments, undermining support for the programs.

State and local budget cuts also are forcing many communities to reassess recycling programs. Several cities, including beleaguered Washington, D.C., dropped residential recycling altogether. Miami's City Council planned to do the same until public protests against the move. Many localities have cut back on frequency of collection and taken other steps to reduce costs. [27]

Even in Washington state, one of the leaders in recycling efforts, budget cuts are taking a toll. The Clean Washington Center is a state agency that helped Seattle and Tacoma achieve recycling rates of about half their waste by developing markets for recyclables. But last year the center, which also provides information to recycling programs throughout the country, was cut from the state budget altogether and forced to become an independent organization.

Outlook

No End of Trash

While recycling programs are not growing as fast as they did in the early 1990s, many experts predict that they will continue to expand. "The amount of trash out there is tending to grow at a slower rate than in the past," Franklin says. "Longer-term, we expect the recycling rate to reach 30 percent of the waste stream by 2000 and 35 percent by 2020." He predicts that most of the increase will result from expanding municipal composting facilities to include more yard waste and food waste rather than new materials collected for sale to manufacturers. "We'll see more composting of paper that's too contaminated for traditional recycling, such as food packages, paper napkins, food waste and even wood."

Consumer preference is likely to be a key determinant of recycling's future in the United States. Americans—both manufacturers and consumers—may embrace recycling on the supply side by participating in recycling programs, but they have been less enthusiastic consumers of recyclable materials and recycled-content end products. "Recyclables tend to be the last hired and the first fired," Keller quips. "If the economy is good, more recyclables get used, but if the economy is not going well, people go back to things they're more comfortable with."

In the absence of pressing consumer demand for recycled products, Raymond predicts, manufacturers are unlikely to boost production of these goods. "We don't have a crisis, virgin materials are cheap, energy is cheap and tip fees are very low, so industry has no incentive to use more recycled materials," Raymond says. "That's just corporate behavior as history has always shown it to be."

Many experts say the decade since curbside recycling took off, boosting the supply of recyclable materials, is too short a time to accurately assess the recycling market's full potential. Technological advances continue to broaden the range of materials that can be recycled cost-effectively, and new processing facilities can be expected to boost demand for post-consumer materials. "We will never reach the point where industries will be dependent on recycled materials alone," Keller says. "Although steel can be recycled an infinite number of times, paper loses a little fiber every time it's recycled, so we're going to still need virgin materials. But the technology will continue to improve, and we'll find ways to recycle materials now considered non-recyclable. There's no question that as technology changes, the variety of things we can recycle will improve over time."

Anticipating such an expansion, the Chicago Board of Trade in 1995 set up an electronic listing for recycled plastic, paper and glass that helps link buyers and sellers of these materials and thus makes the market more efficient. "It's not a full-fledged commodities market," Scarlett says, "but it's growing and provides another tweaking mechanism that helps increase information flows and improve recycling markets."

FOR MORE INFORMATION

U.S. Environmental Protection Agency, Solid Waste and Emergency Response, 401 M St. S.W., Suite 5101, S.E. 360, Washington, D.C. 20460; (800) 424-9346 (in the Washington area, (703) 412-9810); http://www.epa.gov. EPA's Solid Waste Hotline offers information on how to contact recycling coordinators at the local level.

Reason Foundation, 3415 S. Sepulveda Blvd., Suite 400, Los Angeles, Calif. 90003; (310) 391-2245; http://www.reason.org. This nonprofit, nonpartisan public policy think tank advocates market solutions to environmental problems. It has done economic studies of recycling.

Waste Policy Center, 211 Loudoun St. S.W., Leesburg, Va. 20175; (703) 777-9800. This independent consulting and research organization analyzes the costs and benefits of recycling.

Environmental Defense Fund, 257 Park Ave. South, New York, N.Y. 10010; (800) 684-3322; http://www.edf.org. EDF provides information on the economic and environmental benefits of recycling.

Solid Waste Association of North America, P.O. Box 7219, Silver Spring, Md. 20907; (800) 677-9424; http://www.swana.org. Representing government officials who manage municipal solid waste programs, this group provides information on recycling, combustion and other alternatives to waste disposal.

U.S. Conference of Mayors, Municipal Waste Management Association, 1620 I St. N.W., 6th floor, Washington, D.C. 20006; (202) 293-7330. This organization of local governments and private firms involved in waste collection and recycling helps communities plan recycling programs.

Of course, industry responds to consumer demand as well as the cost of raw materials. And for the most part, Americans remain unreliable end users of products made from recycled materials, which many view as inferior in quality.

"People in the food-processing industry will tell you that as a rule putting the green seal of approval on a product indicating it is environmentally benign is going to cost them market share," Taylor says. "People avoid it because they think it's flimsy, unsanitary or not up to standard performance."

Even active participants in recycling programs overlook recycled-content products. "Sure, there are some 'greens' out there, but the vast majority of us are more sensitive to price and quality than to environmental concerns," Raymond says. "Do I have the time to read the labels on all the products that I buy in the grocery store when I have only 30 minutes to get the shopping done? Get real."

Recent trends suggest those attitudes are slowly changing, however. Environmentally concerned consumers are driving growth in retail businesses such as the Fresh Fields-Whole Foods grocery chain, which specializes in environmentally benign products, including those with recycled content. In any case, consumer demand for recycled-content products may not be necessary for further expansion of this market, which Keller says already accounts for at least $10 billion in sales each year, because so many products include recycled materials even though they don't bear the green seal.

"There is no question that there are still remnants of the population who view recycled materials as inferior," Keller says. "But lots of recycled-content products have been quietly used for decades. There is no steel in the United States that doesn't have a minimum of 25 percent recycled content. There are no paper mills being built in the United States today that depend on virgin materials alone. Even *Air Force One* flies on retread tires." ❖

Thought Questions

1. How could the market for recycled materials be stabilized?
2. Should the government legislate that certain products (e.g., toilet tissue) must be made from recycled materials? Why or why not?
3. Are market factors the only factors that should be considered when deciding whether recycling is worthwhile? What other factors should be considered?
4. Do you think that the German system of recycling could work in the United States? Why or why not?
5. How might the recycling of heavy materials such as glass be made more cost effective?

Notes

[1] See Franklin Associates Ltd., *Solid Waste Management at the Crossroads* (December 1997). For background, see "Garbage Crisis," *The CQ Researcher*, March 20, 1992, pp. 241–264.

[2] Franklin Associates Ltd., *op. cit.*, pp. 1–16. Franklin Associates provides the EPA with data on recycling. See also J. Winston Porter, *Trash Facts IV*, Waste Policy Center, 1997.

[3] John Tierney, "Recycling Is Garbage," *The New York Times Magazine*, June 30, 1996.

[4] Richard A. Denison and John F. Ruston, "Recycling Is Not Garbage," *Technology Review*, October 1997.

[5] The poll, conducted by American Opinion Research for the now-defunct Council on Packaging in the Environment in 1996, was cited in Bill Noone, "'Closing the Loop' Remains a Priority for Packagers," *Packaging Technology & Engineering*, November 1996.

[6] Porter, *op. cit.*

[7] Denison and Ruston, *op. cit.*

[8] For background on air pollution, see "New Air Quality Standards," *The CQ Researcher*, March 7, 1997, pp. 193–216.

[9] Lisa A. Skumatz, "Nationwide Diversion Rate Study: Quantitative Effects of Program Choices on Recycling and Green Waste Diversion: Beyond Case Studies," Skumatz Economic Research Associates Inc., July 1996.

[10] Information in this section is based on Frank Ackerman, *Why Do We Recycle?* (1997), pp. 14–19.

[11] For background, see "Cleaning Up Hazardous Wastes," *The CQ Researcher*, Aug. 23, 1996, pp. 745–768 and "Water Quality," *The CQ Researcher*, Feb. 11, 1994, pp. 121–144.

[12] For background, see "Environmental Movement at 25," *The CQ Researcher*, March 31, 1995, pp. 288–311.

[13] For background, see "Renewable Energy," *The CQ Researcher*, Nov. 7, 1997, pp. 961–984.

[14] Ackerman, *op. cit.*, p. 18.

[15] *Ibid*, pp. 11–12.

[16] *Ibid*, p. 18.

[17] See Lynn Scarlett, "Roller Coaster Recycling Markets: Down, Up, and What's Next?" *MSW Management*, January/February 1996, p. 51.

[18] *Ibid.*

[19] See Jessica Lucyshyn and Robert Craggs, "A Five-Year History of Recycling Market Prices: 1997 Update," *Resource Recycling*, February 1998, p. 16.

[20] See Franklin Associates Ltd., *op. cit.*, pp. 4–8.

[21] *Ibid.*, pp. 4–14.

[22] *Ibid.*, pp. 4–12.

[23] "'America Recycles Day' to Raise Awareness of Recycling and Buying Recycled Products," *PR Newswire*, Nov. 14, 1997.

[24] Franklin Associates Ltd., *op. cit.*, pp. 4–9.

[25] See Susan Warren, "Environment: Polyester Trash Is Pure Plastic after an 'Unzip,'" *The Wall Street Journal*, Nov. 6, 1997.

[26] Franklin Associates Ltd., *op. cit.*, pp. 4–9.

[27] See Jim Glenn, "Year End Review of Recycling and Composting," *BioCycle*, December 1997.

Chronology

1940s *The first widespread residential recycling efforts emerge during World War II as consumers contribute scrap metals and paper to wartime industries.*

1960s *Concern builds over the nation's mounting waste stream.*

1962
Author Rachel Carson warns of the dangers of toxic chemicals left in landfills in her best-seller *Silent Spring*.

1965
The Solid Waste Disposal Act launches federal research into the technology of waste disposal and provides technical and financial assistance for state and local waste programs.

1970s *The environmental movement spurs interest in residential recycling.*

1970
The Resource Recovery Act extends federal assistance to recycling programs and waste incineration and requires states to come up with plans for managing their waste.

1975
More than 100 communities across the country have introduced curbside recycling, mostly for newspapers, in an effort to boost recycling. More than a quarter of all aluminum beverage cans, the most valuable post-consumer material, are recycled.

1976
The Resource Conservation and Recovery Act (RCRA) sets national standards for landfills and incinerators. By requiring the use of expensive landfill liners and smokestack scrubbers, the law indirectly encourages recycling.

1978
The Public Utilities Regulatory Policies Act (PURPA) encourages the development of waste-to-energy plants. Support for these alternatives to landfills later wanes, however, because of their emissions of toxic ash.

1980s *Acting on fears that landfill space is running out, communities across the country step up curbside recycling programs.*

1987
The *Mobro 4000*, a garbage barge out of Long Island, attracts national attention as it sails down the East Coast to the Caribbean in the vain search for a place to dump its cargo. The incident prompts concern, largely unfounded, that the entire country faces a landfill crisis.

1988
The Environmental Protection Agency (EPA) sets a recycling goal for the United States at 25 percent of total waste.

1990s *Recycling comes under scrutiny after a series of supply gluts causes upheaval in the markets for recycled materials.*

1991
Manufacturers' demand for recycled materials fails to keep pace with the supply following the sudden increase in curbside programs, causing the first major crash in the market for recyclables.

1993
President Clinton issues an executive order requiring all federal agencies to buy printing and writing paper with at least 30 percent recycled content by the end of 1998.

1995
The nationwide recycling rate reaches 27 percent, exceeding the EPA's 25 percent goal. Prices of recycled newsprint and some other materials reach historic peaks, prompting "garbage rustlers" to raid curbside bins for valuable materials. The Chicago Board of Trade sets up an electronic listing for recycled plastic, paper and glass that helps link buyers and sellers of these materials and thus improves the market's efficiency.

1996
Newsprint prices sink so low that some recycling contractors dump collected paper in landfills.

1997
The pace of new curbside programs begins to slow because most programs already collect the materials that are in demand by manufacturers.

At Issue:

Does recycling make economic sense?

RICHARD A. DENISON AND JOHN F. RUSTON—Denison is a senior scientist and Ruston is an economic analyst at the Environmental Defense Fund

From "Anti-Recycling Myths," Environmental Defense Fund, July 18, 1996.

Recycling is not just an alternative to traditional solid waste disposal, it is the foundation for large, robust manufacturing industries in the United States that use recyclable materials. These businesses are an important part of our economy and provide the market foundation for the entire recycling process. . . .

Recycling provides manufacturing industries with raw materials that are less expensive than virgin sources, a long-term economic advantage that translates into value for consumers who ultimately spend less on products and packaging. For example, in the area of paper manufacturing, new mills making paper for corrugated boxes, newsprint, commercial tissue products and folding cartons have lower capital and operating costs than new mills using virgin wood. . . . Recycling has long been the lower-cost manufacturing option for aluminum smelters, and is essential to the scrap-fired steel "mini-mills" that are part of the rebirth of a globally competitive U.S. steel industry. . . .

In a recent study examining 10 Northeastern states, recycling was found to have added $7.2 billion in value to recovered materials through processing and manufacturing activities. These activities employed approximately 103,000 people, 25 percent of them in materials processing and 75 percent in manufacturing. . . .

Market prices for materials like polystyrene are set in the near term by supply and demand forces, underpinned by a host of production cost factors, many of which have nothing to do with environmental impact. An entire sub-discipline of environmental economics has developed to address a range of environmental damages, called externalities, that are not reflected in market prices even in the most regulated industries. . . . [When] a coastal wetland in the Carolinas is converted to a pine plantation and results in damage to estuarine fish hatcheries or reduced water quality, such impacts are certainly not captured in the market price of wood taken from the site.

Nor are any of the costs of disposal included in product prices. If someone drains motor oil from a car into the gutter, it may pollute surface water or groundwater. But the price originally paid for the oil does not anticipate its proper or improper disposal. Finally, another major obstacle to incorporating environmental factors into market prices is the difficulty or impossibility of assigning a meaningful economic cost to such "goods," for example, the value of preserving a rare animal or plant species.

YES

JERRY TAYLOR—Director of natural resource studies at the Cato Institute and senior editor of Regulation magazine

From "Minimum Content, Minimum Sense," The Cato Institute, April 25, 1997.

Ten years into America's holy war against garbage, the case for residential curbside recycling has run smack into the harsh realities of economics. If resources are indeed becoming more scarce, they have a funny way of showing it. Prices for energy, minerals and paper have continued to fall as they have over the course of the century. . . . Post-consumer material is less competitive with virgin material than ever before. . . .

"But," you might point out, "what about the environmental externalities of the mining, timber, paper and energy industries? If you accounted for that, wouldn't recycling be competitive?" Again, not necessarily. First, we have no reliable means by which we can "price" those externalities. Second, those industries do spend tens of billions of dollars annually to comply with federal and state environmental regulations. Are the environmental externalities they impose greater than, less than or equal to the regulatory costs they pay to do business? No one knows for sure, but a number of respected economists . . . strongly suspect the environmental externalities of those industries are more than paid for through the cost of regulatory compliance.

Nor are the externalities of recycling's alternative all that impressive. EPA regulations now ensure that solid waste landfills cause only one additional cancer risk every 13 years, and that's assuming we use such worst-case scenarios and assumptions that even that figure, according to most risk assessment specialists, probably overestimates the actual risk by 100 to 1,000 times the actual risk. Likewise, municipal waste incinerators, according to those same worst-case assumptions, pose less than a 1 in 1 million risk to neighboring communities. . . .

Finally, recycling has its own environmental externalities that must be put into the equation. After all, the actual process of extracting usable raw material from a product is an industrial activity every bit as involved as the process of combining various raw materials to make a product. Both are industrial activities. And both create waste. For example, recycling 100 tons of old newsprint generates 40 tons of toxic waste. Is this consequential? Sure. EPA has reported that 13 of the 50 worst Superfund sites are/were recycling facilities.

If recycling makes economic sense, we don't need to mandate it. And if it doesn't, we shouldn't. You can make a silk purse out of a sow's ear . . . but it's usually cheaper to use silk.

NO

Bibliography

Selected Sources Used

Books

Ackerman, Frank, *Why Do We Recycle? Markets, Values, and Public Policy*, Island Press, 1997.
The author, a professor at Tufts University's Global Development and Environment Institute, reviews the history and market development of recycling. He argues that environmental as well as economic concerns must be included in any assessment of recycling's value.

Articles

"America's Recyclers: A Funny Sort of Market," *The Economist*, Oct. 18, 1997, pp. 63–64.
Government mandates have skewed the markets for recyclable materials, according to this article, by increasing supply with no concern for demand by industry.

Denison, Richard A., and John F. Ruston, "Recycling Is Not Garbage," *Technology Review*, October 1997.
In this response to an earlier critique of recycling, two researchers at the Environmental Defense Fund point out the economic and environmental benefits of recycling and call for efforts to improve the efficiency of community programs.

Lucyshyn, Jessica, and Robert Craggs, "A Five-Year History of Recycling Market Prices: 1997 Update," *Resource Recycling*, February 1998.
Two recycling analysts with R.W. Beck Inc., a national consulting and engineering firm, point out that markets for recycled materials have become less volatile in the past two years and suggest that program managers may be better prepared to avert the oversupply that buffeted the markets in 1996.

Scarlett, Lynn, "Roller Coaster Recycling Markets: Down, Up, and What's Next?" *MSW Management*, January/February 1996, pp. 50–53.
The gyrations in the markets for recycled materials seen in the mid-1990s are likely to continue, the author writes, because numerous events that are beyond the control of governments will continue to affect demand.

Skumatz, Lisa A., Erin Truitt and John Green, "The State of Variable Rates: Economic Signals Move into the Mainstream," *Resource Recycling*, August 1997, pp. 31–35.
By charging consumers for the amount of non-recyclable waste they generate, the authors write, communities can greatly increase recycling rates. More than 4,400 communities in the United States and Canada have integrated this approach into their waste collection programs.

Reports and Studies

Franklin Associates Ltd., *Solid Waste Management at the Crossroads*, December 1997.
This research firm, which provides recycling data to the EPA, predicts that the recovery of recyclable materials will continue to grow, though more slowly than in the past, reaching 35 percent by 2010.

Raymond Communications Inc., *State Recycling Laws Update*, Year-End Edition 1997.
State lawmakers enacted a total of 70 recycling bills in 1996, out of nearly 200 bills followed by this annual study. Local governments, however, maintained their commitment to recycling.

Scarlett, Lynn, Richard McCann, Robert Anex and Alexander Volokh, *Packaging, Recycling, and Solid Waste*, Reason Public Policy Institute, July 1997.
This study from a think tank in Los Angeles, Calif., examines the economic costs and benefits of recycling and concludes that government mandates for recycling rates or levels of recycled content in finished products are unlikely to help the environment.

U.S. Environmental Protection Agency, *The Consumer's Handbook for Reducing Solid Waste*, August 1992.
This overview of recycling includes a guide to help consumers understand what types of materials are commonly included in curbside programs as well as community drop-off centers.

U.S. Environmental Protection Agency, *Manufacturing from Recyclables: 24 Case Studies of Successful Recycling Enterprises*, February 1995.
Companies specializing in products containing recycled materials are profiled. Most are small manufacturers in or near the communities generating the recyclables. They include makers of paper, plastic, glass and other products.

U.S. Environmental Protection Agency, *Recycling Works! State and Local Solutions to Solid Waste Management Problems*, January 1989.
This dated but still relevant study examines innovative approaches to recycling—including programs that didn't work—in 14 states and communities.

The Next Step

Additional information from UMI's Newspaper and Periodical Abstracts™ database

Environmental Benefits vs. Costs

Barlow, Jim, "Industry Embraces Built-In Recycling," ***Houston Chronicle*****, July 20, 1997, p. E1.**
There's a gradual shift by business into something called life-cycle environmental management. Life-cycle management means companies look at the entire environmental impact of their products—starting from raw material to disposal after its useful life is over.

Fiske, John, "Americans Recycle More Paper Despite Low Demand," ***Christian Science Monitor*****, Oct. 20, 1997, p. 13.**
Americans are recycling more paper than ever before. But for recycling to reduce the amount of virgin materials used and to reduce pollution, more demand is needed for recycled products. The American Forest & Paper Association reports that 63 percent of newsprint was recovered in 1996, up from 43 percent in 1990, to become an important new source of paper fibers. Recycling experts say recovered fiber is becoming as important a source of raw material as virgin fiber—or trees. Recycled fiber now represents 30 percent of the supply of new materials industrywide.

Kendall, Peter, "City Recycling as Much in the Bag as in Landfill," ***Chicago Tribune*****, Dec. 16, 1997, p. 1.**
This fall, officials announced with pride that Chicago's ambitious, controversial and assuredly unique blue-bag recycling program had reached its goal of keeping 25 percent of the city's waste from going into a landfill. The problem is yard waste—and the unsortable dross it creates when mixed with other garbage. The stuff has too much garbage in it to be considered yard waste, which can be tilled into the soil. But it also has too much yard waste in it to be considered garbage, which is sent to the landfill.

Mohl, Bruce, "Wasted effort? The bottle-deposit law works: Four out of every five containers covered by the law get recycled," ***The Boston Globe*****, Nov. 16, 1997, p. BGM23.**
Studies from supporters and opponents of expanding Massachusetts' successful bottle-deposit laws concur—expanding the law would be expensive and inefficient.

Scallan, Matt, "Harahan Leaders Aren't Pleased with Recycling," ***New Orleans Times-Picayune*****, Oct. 10, 1997, p. BK1.**
Harahan started one of the first curbside recycling programs in the New Orleans area in 1991, but six years later several city officials are ready to dump it. City officials say the program collects about 550 tons of recyclables at a cost of about $75,000 a year.

Scallan, Matt, "Recycling Diverts Tiny Portion of Trash," ***New Orleans Times-Picayune*****, Oct. 27, 1997, p. A1.**
Efforts have been made weekly in thousands of households across the New Orleans area for the past few years, yet recycling programs have made only a small dent in the stream of garbage flowing into landfills around the state. With some exceptions, most cities and parishes with curbside programs divert only 8–15 percent of the waste from landfills and into recycling, far short of the 25 percent target set by the Legislature in 1989. Even recycling supporters admit the program is an expensive, "feel-good" solution that has not gone far enough. That kind of arithmetic has some public officials wondering whether, despite its environmental benefits, curbside recycling is worth the money.

Snyder, Mike, "Brown Will Face Issue of Popular, but Costly, Curb Recycling Pickup," Dec. 15, 1997, ***Houston Chronicle*****, p. A17.**
About 17 percent of Houston's residents use the city's curbside recycling operations, which is considerably lower than city officials' original goal of 100 percent. In 1997, the city estimates spending about $2 million to operate the curbside program and about $500,000 for eight drop-off recycling centers.

Volokh, Alexander; Scarlett, Lynn, "Is Recycling Good or Bad—or Both?" ***Consumers' Research Magazine*****, September 1997, p. 14.**
Recycling makes economic and environmental sense in some cases and not in others, but the challenge is to know the difference.

Worthington, Rogers, "Suburban Recycling May Be Near a Peak: Northwest Surpasses 40 percent Rate, But Landfill Capacity a Concern," ***Chicago Tribune*****, March 6, 1997, p. 1D1.**
A decade ago, many suburban officials fretted about brimming landfills and loaded garbage trucks with no place to unload. Then they got serious about recycling. In fact, the northwest suburbs have been so successful in recycling their residential trash that they may have maxed out, said Brooke

Beal, executive director of the Solid Waste Association of Northern Cook County, which has 23 member municipalities. The northwest suburbs also are well ahead of neighboring DuPage and Lake Counties, which each recycle about 30 percent of their garbage.

Market for Recyclables

"Biking Your Way to a Recycling Company," ***Biocycle: Journal of Waste Recycling*****, October 1997, p. 18.**
Graham Bergh's Portland, Ore.-based Resource Revival recycles used bicycle parts into CD racks, picture frames and other household goods.

"Carpet Recycling Project Proposed," ***Biocycle: Journal of Waste Recycling*****, January 1998, p. 22.**
Allied Signal and DSM Chemicals are planning a joint venture that should recycle more than 200 million pounds of used carpet each year. The facility will be located in Augusta, Ga.

"Have Bike, Will Recycle," ***Biocycle: Journal of Waste Recycling*****, August 1997, p. 20.**
The Fresh Aire Delivery Service collects recyclables using bicycles. The company currently collects recyclables from 230 residents and 30 businesses in Ames, Iowa, as well as 160 locations at the Iowa State University campus.

"Institute Encourages Steel Recycling," ***Workbench*****, October 1997, p. 16.**
North America's most recycled consumer product is the automobile, and steel recycling saves enough energy annually to power one-fifth of the homes in the United States. The Steel Recycling Institute hopes to see 25 percent of new homes framed in steel by the year 2000.

"Recycling Textbooks," ***Biocycle: Journal of Waste Recycling*****, November 1997, p. 18.**
Textbook industry executives have formed Book Value Inc. to recycle textbook samples that might otherwise end up in landfills when a proposed curriculum is rejected.

Biddle, David, "What's Wrong with Office Paper?" ***Biocycle: Journal of Waste Recycling*****, November 1997, p. 75.**
Deinking mills have increased the flow of office paper wastes in the recycling stream, but lower grades of paper have been a problem. Recyclers are looking for a higher-quality mix.

Brown, Warren, "Chrysler Close to Turning Plastic Recyclables Into a Car," ***The Washington Post*****, Sept. 9, 1997, p. C3.**
Chrysler takes material from pop bottles, adds chopped glass and something to resist the effects of ultraviolet rays, and puts in impact-resistant material such as rubber. These materials are then formed into a four-piece body that is put on a frame. Such a car could help Chrysler stay well on the right side of federal fuel economy standards in the United States, which says new-car fleets must average 27.5 miles per gallon. The CCV is designed to get 50 miles per gallon.

Carrier, Jim, "U.S. Home going 'green built'," ***The Denver Post*****, March 9, 1997, p. H1.**
Tree-hugging became a mainstream business last week when U.S. Home, a $1 billion contractor, announced that every one of its new homes in the Denver area will be "green built," with more insulation, efficient appliances and recycled materials used in construction. The Houston-based company will build 900 homes in the metro area this year, each one containing environmentally friendly features required by an industry checklist. Their key improvement will be wrapping basement walls with insulation and stuffing insulation into hard-to-reach areas near the roof. U.S. Home's announcement triples the size of the Green Builder Program of the Home Builders Association of Metropolitan Denver. Last year, about 300 homes qualified.

Cortez, Angela, "Coors 'Desperate' for Recycled Glass," ***The Denver Post*****, Jan. 11, 1998, p. C1.**
Although one of the state's largest trash haulers prefers not to pick up curbside glass for recycling anymore, the Coors Brewing Co. says it is "desperate" for glass and is urging people to continue to recycle. Coors spokesman Jon Goldman said recycling glass may not mean big money, but it can turn a profit and it's the responsible thing to do because recycling keeps the glass out of the landfills. The company's move concerned Goldman because Coors depends on recycled glass to make new bottles. Recycling glass also burns less energy than making new glass.

Eaton, John, "At Recycled Products, Plastic Makes Perfect," ***The Denver Post*****, May 5, 1997, p. E1.**
Its products are attracting the attention of heavy hitters such as Sears, Roebuck & Co., Gates Rubber Co. and Payless Cashways' Hugh M. Woods stores. They're looking at the company's PlastiFence and its ParkingSpot. The products, both with patents pending, are unique, says Recycled Products CEO Gene Pendery. Perhaps most significant is what the fence does for the environment. Or what it doesn't do: "Each 100 feet of our fence will save the dumping of 5,000 one-gallon plastic jugs in a landfill," he says.

Ericson, Edward Jr., "Recycling the Army Way," ***E: The Environmental Magazine*****, March 1997, p. 16.**
The Pentagon uses Depleted Uranium (DU) for artillery shells and armor plating for tanks, and it is now being fashioned into bullets. DU is a dense radioactive waste product left over after extracting U-235 to make bombs.

Esparza, Santiago, "Program Lets Cell-Phone Users Recycle Discarded Batteries," ***Detroit News*****, Feb. 13, 1997, p. C8.**
For years, people have thrown away cellular phone batteries that could no longer be recharged. The batteries, made of nickel and cadmium, would then sit in landfills adding to the toxicity of the dump. Now the environmentally conscious can drop the batteries off at national retail stores, or even set up their own collection site. Collected batteries are sent to a nonprofit corporation that breaks batteries down, using the nickel for stainless steel and cadmium for making new cellular phone batteries.

Grogan, Peter L., "No Miracle Markets," ***Biocycle: Journal of Waste Recycling*****, January 1998, p. 86.**
The growth of the middle class in Asia is creating a huge market for American recyclable commodities. Grogan discusses this growing market and how the United States can make the most of it.

Pardo, Steve, "In Fraser: Duo Arrested in Recycling Scheme," ***Detroit News*****, Oct. 21, 1997, p. D3.**
Two Pennsylvania men were arrested at a Meijer store in Fraser after they brought at least 20,000 out-of-state cans to turn in for Michigan's dime deposit. They are the second pair from outside Michigan to be arrested this month at the same store on charges they tried to take advantage of the law that pays 10 cents per used bottle or can—the highest rate in the nation. The two men bought $500 worth of scrap aluminum cans, put them into a trailer and drove the cargo to Michigan, police said.

Government's Role

"Mayor Argues Asphalt, Concrete Count Toward Recycling Goals," ***Biocycle: Journal of Waste Recycling*****, August 1997, p. 23.**
The State Supreme Court will rule on New York City's appeal of a decision that asphalt and concrete do not count toward recycling goals. The city has fought recycling quotas in court before and lost, but the city has delayed the deadline for meeting the quota of recycling one-quarter of the city's waste, which it does not meet if asphalt and concrete are not counted.

Bukro, Casey; Young, David, "EPA Ruling on Scrap Gives Recycling a Boost," ***Chicago Tribune*****, May 22, 1997, p. 3.**
A new EPA ruling says scrap is not waste. After years of campaigning by the scrap industry, EPA Administrator Carol Browner on April 18 signed a final rule-making decree declaring that scrap metal is not solid waste under federal law. The ruling is especially important to metropolitan Chicago and the Midwest because both are major recycling centers.

Howe, Peter J., "State Eyes a Broader 'Recycle or Pay' Policy," ***The Boston Globe*****, Sept. 22, 1997, p. A1.**
Massachusetts state officials are proposing to offer communities cash bounties for every ton they recycle and pay for the cost of launching "pay-as-you-throw" trash collections.

Pillsbury, Hope, "Standardizing Recycling Measurements," ***Biocycle: Journal of Waste Recycling*****, January 1998, p. 39.**
After several years of study, experimentation and field testing, the EPA has come up with a standardized methodology for measuring recycling rates. Pillsbury discusses these new standards and what they mean.

Schroer, Bill, "Recycling Program Generates Capital Where Drivers' Rubber Meets the Road," ***The Denver Post*****, March 10, 1997, p. C4.**
Since the late 1960s, the environmental movement has earnestly embraced recycling by connecting consumption to mining, resource depletion and groundwater pollution. In 1993, the Colorado Legislature, with an understanding of these conditions and the vast public appeal for recycling, established Renew at the Colorado Housing and Finance Authority. This agency is a state-government-created, but independent, authority that provides housing and economic-development financing. Renew offers private or nonprofit recyclers, or firms that incorporate waste-diversion activities or recycling in their operations, below-prime-rate loans for working capital, equipment or real estate. Because applicants need not be recyclers to seek these loans, thousands of Colorado businesses are eligible for this financial assistance.

Scott, Peter, "Walton to Begin Quantity-Based Garbage Fees," ***Atlanta Constitution*****, Jan. 15, 1998, p. XJR7.**
This article describes Walton County, Ga.'s "pay-as-you-throw" system of trash hauling. If citizens don't wish to utilize one of six designated sites throughout the county, they have the option of hiring a private hauler.

Landfills

"City Exceeds Its Recycling Goal of Diverting Garbage from Landfills," ***Chicago Tribune*****, Sept. 29, 1997, p. 2C3.**

Chicago's recycling efforts are going so well that the city says it has exceeded its goal of diverting 25 percent of all city-collected garbage from landfills.

"More Is Recycled, but There's Also More of It," ***Christian Science Monitor*****, July 21, 1997, p. 16.**

It may be garbage to you, but it's 'municipal solid waste' to trash professionals. Here are highlights from a report on municipal solid waste prepared for the EPA and released last month: 208 million tons of municipal solid waste were generated in the United States in 1995, down 1 million tons from 1994. (In 1960, it was 88 million tons.)

Americans love living near the beach. Fifty-five percent of the population lives within 80 miles of a major coastline, and that number is projected to increase to 75% by 2025. Coastal communities enjoy many economic benefits associated with recreation and tourism, fishing and shell-fishing, and ship-borne commerce, yet they are also subject to natural processes that can be hazardous to life and property. Currents moving parallel to shore transport sand along the coasts, causing barrier islands to migrate (see Press & Siever, pp. 383–390; Merritts, De Wet, & Menking, pp. 312–315, for a discussion on coastal erosion). Structures once in the middle of an island find the sand slipping away around them and may eventually be swallowed by the sea unless they can be relocated further inland. Winter storms remove sand from beaches and store it offshore in large underwater bars. As a result, high waves encroach further inland where they threaten beachfront properties. Human activities likewise lead to coastal hazards. In an attempt to keep migrating sandbars from blocking harbors, many communities have built jetties to deflect sand to deeper water offshore. These structures interrupt the normal transport of sand along the coast and cut off the supply to the beach down current from the jetty. Beaches begin to erode, allowing waves greater access to coastal properties. In an attempt to halt further beach erosion, communities then construct groins—concrete and rock structures oriented perpendicular to the beach designed to trap migrating sand. They also make concrete or wood sea walls, and place rip-rap (boulders) against eroding shores to try to stabilize the beach environment. All of this construction has diminished the natural beauty of the coasts, and has led many to question the wisdom of coastal development.

Another method of mitigating erosion, beach replenishment, does not lead to an unsightly coast, but is very expensive. Dredges remove sand from offshore and spray it into the nearshore to widen the beachfront. Communities like Ocean City, Maryland, depend on this process to protect them from the high waves generated by storms. The federal government has pledged $500 million to keep Ocean City's beach in place over the next 50 years. That the taxpayer is footing the bill for beach replenishment and other methods of coastal stabilization has angered many who feel that their hard-earned dollars are being used to subsidize wealthy communities. On the other hand, officials in towns such as Ocean City point out that the tourism-supported tax revenues they generate far outweigh the expenditures used on stabilizing the shore. Although this may be true for the short term, coastal development is under threat from hurricanes and other large storms capable of resulting in billions of dollars in damage. When disasters such as these strike, the Federal Emergency Management Agency springs into action to assist people whose homes and businesses have been destroyed. The agency employs a combination of grants, low interest loans, and insurance payments to help people rebuild their homes and their lives. Critics of the agency claim that giving this aid only encourages people to redevelop areas known to be hazardous. They also point out that the National Flood Insurance Program, which was designed to shift the responsibility of disaster relief from the federal government to homeowners and businesses, does not collect enough in premiums to pay for itself. Global warming promises to amplify coastal hazards in the future as rising sea levels inundate low-lying areas and as warmer oceans spawn greater numbers of hurricanes (for discussions on sea- level rise, see Press & Siever, pp. 349, 388–390, 564; Merritts, De Wet, & Menking, pp. 308–312). Thus, taxpayers face the possibility of ever-mounting bills to bail out storm-ravaged coastal communities unless development is limited or those communities are left to fend for themselves after disasters.

Coastal Development

Does it put precious lands at risk?

By early in the next century, 75 percent of all Americans will live within 80 miles of an ocean or the Great Lakes. The lure of living close to water has spurred explosive growth in resorts from Ocean City, Md., to North Carolina's Outer Banks. But most of the building is on fragile spits of land prone to washing away in major storms. That increases the chance of a catastrophic loss of life and a multibillion-dollar disaster bailout if a hurricane or huge storm strikes. Many critics are questioning whether federal shoreline-protection policies are encouraging irresponsible growth and leading to other problems like pollution and depletion of fisheries. The Clinton administration is trying to trim some shoreline subsidies but is encountering fierce resistance from coastal state lawmakers in Congress.

Hurricane Opal sheared off a dune in 1995, leaving this beachfront house in Florida perilously close to the water. (AP/Bill Kaczor)

THE ISSUES

29
- Are taxpayers subsidizing high-risk coastline development?
- Does beach replenishment work?
- Should there be a national coastal-management plan?

BACKGROUND

38 **Vacation Boom**
Early in the century, developers began building on narrow barrier islands to accommodate vacationers.

40 **Action by Congress**
Laws in the 1980s discouraged coastal development and set cost-sharing levels for shore-protection work.

41 **Rising Sea Levels**
Increases caused by global warming further threaten new and existing coastal development.

CURRENT SITUATION

42 **Budget Battle**
The Clinton administration and Congress appear headed for another fight over funding for shore protection.

43 **Pressure on Congress**
Fights are brewing over the Law of the Sea treaty and 1.2 million undeveloped coastal acres.

OUTLOOK

44 **Ending the Cycle**
Flood-mitigation efforts seek to end the disaster-rebuild-disaster cycle.

SIDEBARS & GRAPHICS

30 **The Trouble With Sea Walls**
Why they can cause more erosion.

32 **Win Some, Lose Some**
Shorefront communities have had mixed results dealing with erosion.

35 **Corps' Projects Cost More Than $3 Billion**
The Army Corps of Engineers has handled more than 1,300 beach projects since World War II.

37 **The Dreaded Northeaster**
The vicious winter storms make hurricanes seem tame.

39 **To Save a Lighthouse**
The National Park Service wants to move the 1870 Cape Hatteras light.

44 **Mystery of the 'Cell From Hell'**
A toxic microbe symbolizes the new kind of pollution plaguing coastal waterways.

48 **Chronology**
Key events since 1900.

49 **At Issue**
Should the federal government continue to subsidize beach-replenishment efforts?

FOR FURTHER RESEARCH

50 **Bibliography**
Selected sources used.

51 **The Next Step**
Additional sources from current periodicals.

Note: For more information on this topic, please see the following pages in Press and Siever's *Understanding Earth,* Third Edition: pp. 349, 383–390, 564; and in Merritts, De Wet, and Menking's *Environmental Geology:* pp. 308–315.

Coastal Development

By Adriel Bettelheim

The Issues

Off the coast of Ocean City, Md., engineers are making yet another attempt at reversing the forces of nature. Operators of a large dredge suck sand from the ocean floor and shoot it through a pipeline onto beaches that have been eaten away by storms and rising seas.

Two northeasters last winter washed away portions of the popular resort's beach, which had been replenished over a decade at a cost to state and federal taxpayers of $80 million.* But like many storm-battered coastal communities, Ocean City is optimistic about saving its famous strand. The federal government plans to keep pumping sand onto Ocean City's 10 miles of shoreline over the next 50 years at an estimated cost of $500 million—or $1 million a mile per year.

Local officials defend the project, saying the beach protects shoreline properties from erosion and storm damage and attracts tourism that generates some $67 million in federal tax revenues each year. "I hear the beach replenishment being referred to as a subsidy. If it's delivering a 600 percent return, exactly who's subsidizing whom?" asks City Engineer Terry McGean.

But others wonder whether beach projects like Ocean City's are really damaging fragile coastlines rather than helping them. Critics say long-term federal commitments to rebuild beaches, erect erosion barriers and pay flood insurance encourage developers and wealthy individuals to build more properties on fragile spits of land that are at the greatest risk of washing away in major storms. That increases the chance of a catastrophic loss of life and a multibillion-dollar disaster bailout if a hurricane or major storm strikes.

"This kind of an arrangement doesn't exactly force a beach community to make wise long-term decisions," says Stephen Leatherman, a coastal geologist at Florida International University (FIU) in Miami. "These folks build and build, and are capable of forgetting they're putting static structures in the way of a dynamic landscape."

The debate is getting increased attention during this year's series of events marking the United Nations' "Year of the Ocean." At the National Ocean Conference in Monterey, Calif., in June, conferees discussed developing a national plan to deal with development along America's approximately 95,000 miles of shoreline, as well as pollution and other growth-related issues. In 1967, the last time such a plan was discussed, a White House conference known as the Stratton Commission led to the creation of the National Oceanic and Atmospheric Administration (NOAA), the ocean research and weather-forecasting branch of the U.S. Department of Commerce.

The current concerns are driven by Americans' continued attraction to the shore. Approximately 55 percent of the U.S. population—about 145 million people—now live within 80 miles of an ocean coast or one of the Great Lakes. By the year 2025, NOAA estimates close to 75 percent of the population will live in coastal areas. [1]

Many of the most popular tourist beaches and developments from Maine to Texas sit on low-lying barrier islands

*Northeasters, known to mariners as nor'easters, are major storms that occur along the Mid-Atlantic coast in the fall, winter and spring. They are so named because their heavy winds usually originate from the Northeast.

The Trouble With Sea Walls

Structures built too close to the shoreline typically require sea-wall construction to prevent erosion (1). But walls create a barrier to natural beach replenishment, leading to a narrowing of the beach and a steepening of the offshore slope (2). Eventually, the beach disappears, and the increased wave power due to the steepened slope causes the bulkhead to fail (3). Construction of a higher wall leads to more powerful waves and further erosion (4).

1. Before the wall

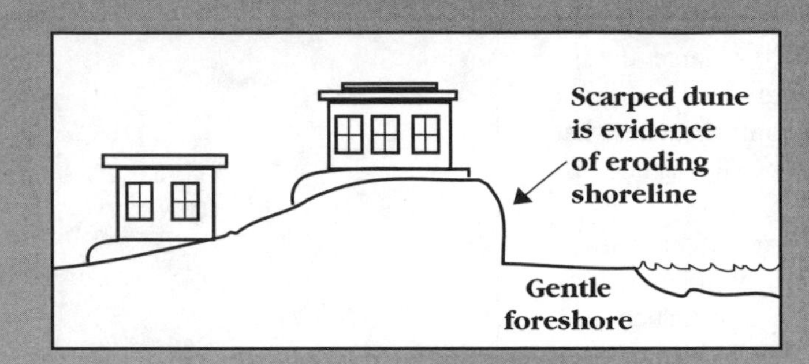

2. Wall constructed; development proceeds as buyers believe property is protected by the wall

3. Two to 40 years later

4. Ten to 60 years later

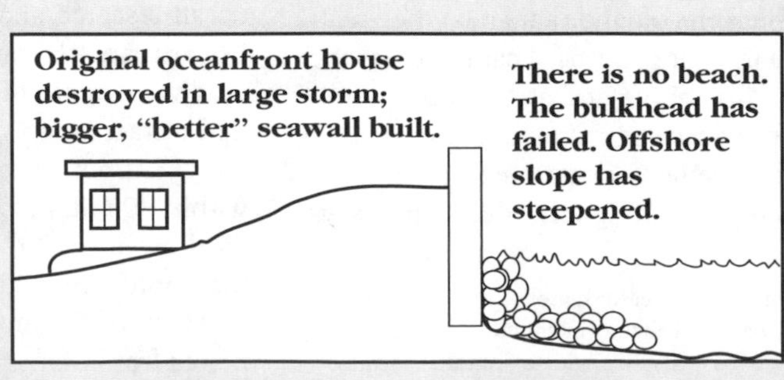

Source: David Bush, Orrin Pilkey and William Neal, Living By the Rules of the Sea, *Duke University Press, 1996*

that constantly migrate, change shape and even fold over on themselves due to tides, storms and sand flow. Anxious property and business owners constantly clamor for more beach fill and breakwaters to maintain a buffer zone between themselves and the sea.

The situation isn't confined to the Atlantic and Gulf coasts: In California and the Pacific Northwest, beach communities are seeking federal aid to stop erosion that reached rates of 10 feet a year during this year's El Niño-driven storms. Generally the problem has not been as widespread on the West Coast because the shoreline is characterized by sheer bluffs, not extensive sandy beaches, and there is less private ownership of coastal land.

Environmentalists argue that continued development—not erosion or bad weather—is the greatest threat to coastlines. They say construction of sea walls, breakwaters and jetties to maintain navigation inlets and protect properties creates obstacles that block the natural flow of sand. That keeps adjacent beaches from being replenished, leaving thin buffer zones that can be easily breached in major storms. [2]

In addition, coastal developments annually discharge approximately 2.3 trillion gallons of partially treated sewage into nearby waters. Nutrients and pathogens in the effluent can foul ecosystems, forcing closure of shellfish beds and swimming areas. Runoff from city streets and parking lots, along with pesticides and fertilizers that leach off farmlands, further threaten water quality. *(See story, p. 44.)*

"It's as if the powers that be develop the shoreline, draw a line in the sand and challenge the sea not to cross it. They want to maintain the line at any cost," says D.W. Bennett, president of the American Littoral Society, an environmental advocacy group in Sandy Hook, N.J. "Towns, counties and states won't seriously think about alternatives to building as long as the federal government assures it's going to be a free lunch."

Charting a course to address all the concerns is proving difficult. U.S. coastal policy is a patchwork of regulations administered by more than a dozen federal agencies that sometimes have conflicting agendas. Since Congress established the Beach Erosion Board in 1930 to study deteriorating coastlines, legislation has increasingly made the federal government responsible for protecting private shorefront property, preventing floods, preserving critical coastline habitats and working with states to define hazard zones. Such a broad agenda has led to bureaucratic wrangling and lawsuits.

In one oft-cited example, local fishermen, the U.S. Army Corps of Engineers and environmentalists have skirmished since 1970 over a plan to build twin, mile-long jetties at Oregon Inlet, N.C., north of Cape Hatteras. Fishing interests say channels within the inlet are migrating with tides and storms, hindering navigation. Congress authorized construction of the jetties but never funded the project. The U.S. Department of the Interior, which owns land adjacent to the inlet, opposes the project, saying construction will endanger wildlife refuges and possibly exacerbate erosion. The corps, which designs the nation's shore protection and flood control projects, maintains the jetties will at last stabilize the inlet and eliminate the need for continually dredging navigational channels.

Scientists estimate 80–90 percent of the U.S. coastline is going the way of Oregon Inlet, eroding at rates varying from one to two feet a year to upward of 14 feet a year. They question whether the government should restrict development, move the most endangered properties away from beaches and allow coastlines to re-form naturally over time.

The Clinton administration is taking a more measured approach, attempting for the past four years to selectively reduce shoreline subsidies. In its 1999 budget proposal, the White House called for trimming the federal government's share of long-term beach replenishment and slashing the corps' shore-protection budget by 79 percent from the 1998 level of $108.9 million.

That isn't sitting well with coastal state lawmakers, who have ignored the administration's recommendations and voted to keep funding many of the beach projects. The lawmakers say cuts would threaten beaches that generate hundreds of millions of dollars in annual revenues for coastal communities and regional economies.

"Beachfront communities represent one of the most thriving sectors of the nation's economy," says Sen. Frank R. Lautenberg, D-N.J. "Stop coastal protection, and you're letting a lot of money wash out to sea."

Adds Rep. E. Clay Shaw Jr., R-Fla., "It's much more than saving some rich person's front yard."

As coastal interests and development critics grapple, here are some of the questions they are asking:

Are taxpayers subsidizing high-risk coastline development?

When Hurricanes Bertha and Fran slammed into North Carolina's Outer Banks within two months of each other in 1996, the federal government responded with speed and zeal.

On Topsail Island, Shara and Ronald Sullivan of Newton Falls, Ohio, found their $260,000 oceanfront vacation home destroyed. Federal flood insurance covered $121,000 in combined losses from the two storms. But because the couple's lot was too eroded to rebuild, the Small Business Administration (SBA) also chipped in with a $115,000 loan to buy another house one block away. The

Win Some, Lose Some

Shorefront communities have had mixed success dealing with development and beach erosion. Here are some examples:

Presque Isle, Pa.—The Army Corps of Engineers spent millions of dollars trying to stabilize this scenic sandspit in Lake Erie, but flooding and erosion persist. The latest plan calls for installing 58 segmented breakwaters at a cost of $30 million.

Southampton and Westhampton, Long Island—Dozens of exclusive houses built along eroding beaches have been swept into the sea by storms since the 1960s. Southampton officials are debating whether to allow owners to build sea walls to protect their residences.

Sandy Hook, N.J.—The Corps of Engineers plans to spend $15 million next year to replenish about 10 miles of beaches stretching to Manasquan. It's part of a large-scale proposal to rebuild 33 miles of the state's coastline.

Ocean City, Md.—The resort community lost portions of its $80 million replenished beach during last winter's storms. The corps estimates that maintaining the town's 10 miles of beach will cost $500 million over the next 50 years.

Oregon Inlet, N.C.—Watermen and environmentalists are at odds over a plan to build twin jetties to keep the passage open to the Atlantic Ocean. Congress authorized the $100 million project in 1970 but never funded it.

Cape Hatteras, N.C.—Scientists expect the sea to swallow the area's landmark 200-foot-tall lighthouse within two decades. The National Park Service wants to move the 2,800-ton structure away from the eroding coastline.

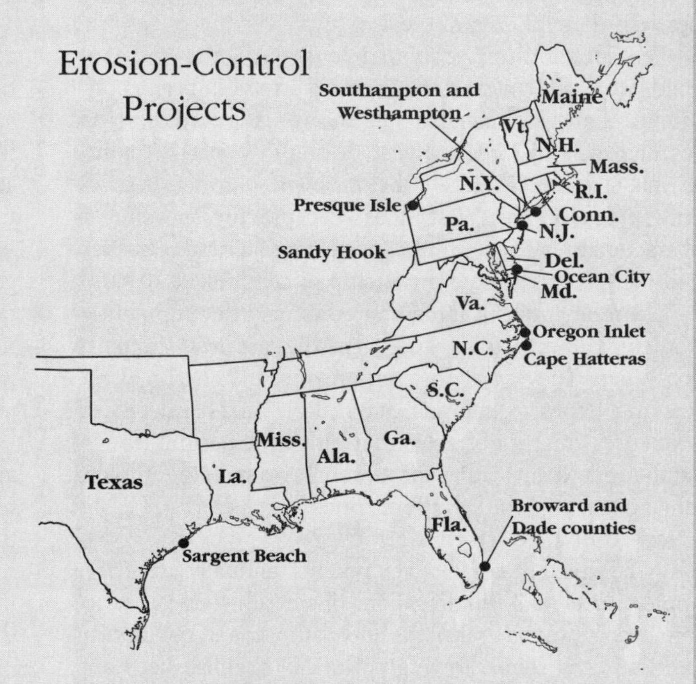

Broward and Dade counties, Fla.—Storms over the last two years have swept hundreds of thousands of cubic yards of sand from South Florida's palm-lined coast. Concerned about the effects on tourism, state officials are pressing for federal beach-restoration help.

Sargent Beach, Texas—Federal officials recently completed an $80 million, 8-mile-long sea wall to prevent the Gulf of Mexico from breaching the Gulf Intra-coastal Waterway. Dredging and jetty construction over the course of a century increased erosion rates.

Sources: Orrin Pilkey and Katherine Dixon, The Corps and the Shore, International Hurricane Center, Florida International University; news reports.

couple acknowledges the risk of living on the erosion-prone island.

"I don't think I'd take the risk if I couldn't get some kind of insurance," said Shara Sullivan. "I'm too chicken for that." [3]

The Sullivans' payments were part of $1 billion in grants, loans and insurance claims the federal government paid out to North Carolinians within a year of the hurricanes. To many, the payments symbolized Washington's increased sensitivity to the perils of living along the coast—and an attempt to avoid the criticism for slow response that occurred after Hurricane Andrew hit South Florida and Louisiana in 1992.

But scientists and government officials have increasingly questioned why the government is paying to rebuild beach homes, condominiums and other developments in high-risk areas that private insurers are reluctant to cover. And they are more closely scrutinizing the way billions of dollars in disaster relief are spent.

Much of the criticism is directed at the National Flood Insurance Program, established in 1968 to cover flood-prone shore and inland communities. Skeptics say the program's subsidized insurance discourages property owners from exercising personal responsibility and building in safer areas. Premiums average about $300 nationwide for $100,000 in coverage. The maximum coverage available for a single-family home, not including contents, is $250,000.

"The program has become a blank check signed by the federal taxpayer: a check whose total, when erosion, rising sea levels, storms and hurricanes take their toll, will be in the billions," environmentalist Beth Millemann wrote in a scathing critique of the program. [4]

The Federal Emergency Management Agency (FEMA), which administers the program, says it actually discourages risky development because flood insurance is a condition for any federally insured mortgage or construction loan in a flood-hazard area. Without such requirements, communities could be as careless with new development as they wished, according to FEMA Director James Lee Witt.

"As more people buy flood insurance, fewer flood victims must be bailed out with tax-funded federal disaster aid," Witt wrote. [5]

Since the late 1970s, FEMA has required coastal policyholders who rebuild after floods to adhere to tougher standards, such as elevating structures to above the 100-year flood level. In 1995, FEMA also embarked on a campaign to sign up homeowners in flood-prone areas who don't currently have insurance—an effort to discourage those who think they would automatically qualify for aid when a flood strikes.

David Conrad, water-resources specialist for the National Wildlife Federation, says the program still has fundamental flaws. He notes FEMA doesn't consider whether a coastline is eroding when it offers insurance in a shore community, instead relying on the more vague historical likelihood of flooding in an area. Also, Conrad says, properties built before FEMA issued flood insurance rate maps in the early 1970s are "grandfathered" into the program, even though they often were built to weaker construction standards and are at greatest risk of being damaged.

The result, critics say, is that the flood-insurance program isn't financially stable or self-sufficient. A series of major disasters—from Hurricane Andrew to the Midwest floods of 1993 to the twin hurricanes of 1996 to last year's Red River floods that devastated Grand Forks, N.D.—triggered massive claims that dwarfed the premiums taken in. The program has had to borrow funds from the U.S. Treasury and pay back the loans with interest out of future premiums. Currently, it owes the Treasury more than $720 million.

Meanwhile, FEMA continues to pay to rebuild structures in high-risk areas. On Topsail Island, owners of 217 properties that have been flooded two or more times have collected $10.9 million from FEMA in national flood insurance payments. Ironically, some properties were in a zone designated off-limits to new coastal development by federal legislation passed in 1982 to preserve fragile coastlines.* But FEMA was still able to send aid by using a loophole that allowed payments when lives or existing properties are threatened.

In California, the agency is paying $1.5 million to rebuild a sea wall to protect a row of at-risk homes in Pacifica, south of San Francisco. The affluent neighborhood's previous sea wall was destroyed by coastal erosion, and El Niño-inspired storms ate away at seaside bluffs under nine homes early this year.

FEMA's inspector general and the U.S. General Accounting Office (GAO), the watchdog arm of Congress, have each recommended that FEMA improve its ability to identify applicants who should purchase flood insurance and develop a database to track compliance.

Witt, who has generally earned high marks for turning around FEMA's relief efforts, says the agency plans to "reengineer" its public-assistance program to avoid encouraging more risky development.

"If we're going to keep people out of harm's way, and if we're going to cut costs from disasters, we're going to have to change the way we do business," Witt said late last year. [6]

FEMA isn't the only agency to draw criticism. After Hurricane Fran, the corps built a 15-mile-long, $4.6 million sand dune to protect Topsail Island property that drew criticism from other federal agencies after it quickly started blowing away. The corps maintains the dune stopped further property damage.

On nearby Wrightsville Beach, the SBA provided more than $1 million in grants to build a temporary sea wall and make improvements at the $22 million Shell Island Resort, a high-rise condominium complex on an eroded spit of land that soon is expected to fall into the sea. North Carolina officials have denied permission to the buildings' owner to erect a permanent sea wall and ordered that the temporary one come down next year. The case is now in court.

---

*The legislation is the Coastal Barrier Resource Act, also known as COBRA.

Critics say the actions of the individual agencies reflect the broader political realities of disaster relief. Ever since the sluggish bureaucratic response to Hurricane Andrew embarrassed the Bush administration in the 1992 election year, the Clinton administration has taken pains to show compassion and speed assistance to stricken communities. Clinton and Vice President Al Gore have visited some of the harder-hit communities and frequently waived cost-sharing requirements with states and local governments, meaning the federal government pays 100 percent of some costs.

White House officials say the practice harks back to Clinton's experience as governor of Arkansas and his sensitivity to the effects disasters have on individuals and communities. Others say guarantees of disaster aid score big points and can secure votes in key states.

"The Clinton administration, more so than any of their predecessors, has realized the political value of disaster spending and being there to hand out the checks," says David DeSanti, executive director of the Natural Disaster Coalition, an insurers' group concerned with rising disaster costs.

Some coastal interests see it as more of a contractual obligation, viewing the aid as part of an unwritten covenant the government entered into when it began building more roads and bridges to the shore as part of the post-World War II construction boom.

"We spend billions of dollars constructing transportation corridors to enable people to get to the beach. Yet, to spend several million on ensuring that the beach exists and to protect the natural and economic resources which justify the transit projects is characterized as extravagant," says Tony MacDonald, executive director of the Coastal States Organization, a Washington lobbying group representing coastal states, commonwealths and territories. "The shortsightedness of such views should be plain."

Does beach replenishment work?

In the late 1970s, Miami Beach was hardly a beach at all. Sea walls built by hotel owners in the 1960s to halt the advancing waves interrupted the natural flow of sand. At high tide, waves lapped up against the concrete barriers, engulfing the famous strand where James Bond first met Goldfinger.

In 1982, the corps completed a massive beach replenishment, pumping 13 million cubic yards of sand that created a 300-foot-wide beach at a cost of about $64 million. Tourist visits soon surged, reviving the moribund city economy. Nearly two decades later, most of the beach remains and officials estimate federal tax revenues from foreign visitors alone exceed $130 million a year.

"It basically rejuvenated the place. Without the beach, it wasn't a place you'd want to visit," says James Houston, a physicist at the corps' waterways experiment station in Vicksburg, Miss., which designs beach-replenishment projects.

The Miami Beach experience is often cited by coastal-development interests as justification for authorizing large beach-fill projects to combat erosion and create tourism.

The Morris Island (S.C.) lighthouse now stands about 2,000 feet at sea. Nearby jetties disrupted sand flow and increased erosion, causing the shoreline to retreat from under the tower in the mid-1940s.

Program for the Study of Developed Shorelines

However, scientists say it may be more of an exception to the rule. The unique shape of Miami Beach's sand grains, made from coral shell fragments, made the beach tightly packed and more resistant to wave erosion. The presence of offshore reefs near the Florida coast also lessened wave action.

More typical, the scientists say, may be Mid-Atlantic beach projects that need to be replenished every three to five years because rising seas and wave action wash away most of the sand. Critics contend this leads to long-term financial commitments—such as the one in Ocean City, Md.—and a continuous cycle of federal beach subsidies.

Understanding how beaches wash away requires looking beyond the visible beach. Wave action constantly scours sand off a beach, depositing it in offshore bars and later redistributing it to the beach. Erosion takes place when storms and winds create waves that remove more sand than they give back, or when people build objects that block the flow of sand.

States and local communities typically seek out the corps for advice when they have erosion problems. The corps conducts feasibility studies and cost analyses that have to meet the approval of Congress. But critics contend this process amounts to writing a blank check; the longer the recommended life of a project, the more money the corps stands to receive in its construction budget. And coastal lawmakers have an incentive to deliver beach-fill projects in much the same way that other lawmakers vie for federal highway aid for their districts.

"Beach nourishment has been oversold by the Army Corps of Engineers," says FIU geologist Leatherman. "They need projects to work on, and the local communities see this constant flow of federal money. It's no surprise it's billed as a panacea for the erosion problem."

"If you lift up the flap, there's more to predicting how the beach will behave than meets the eye," says Orrin Pilkey, a coastal geologist at Duke University in Durham,

Corps of Engineers Beach Projects Cost More Than $3 Billion

Since the end of World War II, the Army Corps of Engineers has undertaken more than 1,300 federally funded beach-replenishment projects costing more than $3 billion. Areas with the most projects are listed below.

	Number of projects	Sand volume (million cubic yards)	Total cost (in $millions)
Florida (East Coast)	144	86.3	$443.2
New Jersey	124	57.4	312.7
Florida (West Coast)	113	46.4	224.8
North Carolina	108	43.5	146.2
Massachusetts	81	3.7	56.4
New York	73	98.2	523.1
Virginia	48	13.6	78.8
Connecticut	44	5.3	48.3
Lake Michigan	280	13	100.8
Lake Erie	54	9.4	77.9
Lake Superior	53	1.4	9.6

Note: Total costs are adjusted for inflation to 1996 dollars and include both known costs and estimates for projects with unknown costs.

Source: Duke University, Center for the Study of Developed Shorelines, June 9, 1998

Development on narrow barrier islands has property and business owners in communities like Ocean City, Md. clamoring for more beach fill to maintain a buffer zone between themselves and the sea.

Adriel Bettelheim

N.C., and a vocal critic of beach fill. "But the communities just assume it's going to last and increase the density of shorefront buildings, which means there will be much more damage if there's a big storm."

Pilkey estimates the corps has spent $3.5 billion over the past 30 years on 1,305 beach replenishments. He notes, however, that the corps uses unreliable mathematical models to calculate the useful life of beach fill. Among the notable failures: the $12 million replenishment of Folly Beach, S.C., in 1993, most of which disappeared within two years without the passage of a significant storm. [7]

Pilkey and other critics also say that the corps' cost estimates are overly optimistic. Northeasters that struck Ocean City, Md., in 1991 and 1992 were calculated to be the type of storms that occur once every 15 years. However, they added about $12 million more to the sand pumping cost than the Corps of Engineers had estimated. [8]

Corps officials defend their work, saying replenished beaches are never guaranteed to stay in place but protect billions of dollars' worth of coastline property that otherwise would get swallowed by the sea. They dispute Pilkey's 30-year cost estimates for beach replenishment, saying expenditures are much lower and now total about $150 million a year. As for the general wisdom of building at water's edge, corps officials say they don't control development, they just save it.

"We get blamed for encouraging development, but people are going to the coast anyway, and they don't even know where the corps' projects are," says Harry Shoudy, a senior policy analyst with the corps in Washington. "It boils down to whether you want to protect them, or not."

Despite the Clinton administration's pleas for less sand pumping, the corps is studying or is in the midst of several mega-projects that continue to stir debate. The largest is designed to maintain a 100-foot-wide beach for 50 years on a 33-mile stretch between Sandy Hook and Barnegat Inlet, N.J., primarily to protect beachfront homeowners from coastal storms. Nearly $100 million in federal funds have been spent so far, and the agency estimates the total cost of the first 21 miles to be $1.1 billion. [9]

On Fire Island, off New York's Long Island, the corps is studying a long-term plan to pump sand to protect 4,000 property owners from storms and erosion. Environmentalists oppose the plan because it may upset the barrier island's delicate ecosystem and disrupt the natural east-west sand drift.

"The biblical adage that one should not build a house on sand is given ultimate demonstration at Fire Island," says New York state Assemblyman Steven Englebright, D-Setauket, who is also a geologist.

But property owners say a "naturalist" approach of letting beaches re-form themselves doesn't correspond with reality. "This is between purists in the environmental movement and those who live in the real world," Fire Island homeowner Ken Entler said at a public hearing last December. [10]

Should there be a national coastal-management plan?

To some, the story of erosion control at Long Island's Westhampton Beach illustrates the need for better coordination among federal, state and local officials on coastal-development problems.

The Dreaded Northeaster

When it comes to coastal weather, hurricanes get most of the headlines. Slow to form and easy to track with radar, they threaten Atlantic and Gulf coast shore communities in late summer with such frequency that weather forecasters use names to distinguish them.

Less recognizable, but arguably more destructive, are northeasters. These giant wintertime storms form quickly and behave unpredictably, born of clashing air masses and strong jet streams. The strong winds, heavy snowfall and flooding they bring inflict some of the most serious weather-related damage to coastlines, stripping away tons of sand and damaging shorefront homes and businesses.

The storms typically begin when a cold-air mass circling clockwise around a high-pressure center moves over the Atlantic coast, meeting a low-pressure system circling counterclockwise. The spinning air masses mesh, picking up warm, moist air from the Gulf coast that rises, cools and condenses into rain or snow. The turbulence sometimes is exacerbated by a second high-altitude, low-pressure system that sucks more air upward, increasing precipitation and winds. [1] The result is a huge storm with a counterclockwise rotating air mass that resembles a hurricane, packing winds from the northeast. The storms typically travel 30–50 miles per hour but don't follow a predictable direction like most hurricanes because the hot and cold air masses clash in different ways.

Such a weather system caused what meteorologists and coastal geologists regard as the worst single storm of the 20th century, the "Ash Wednesday storm" of March 1962. The storm hit the Mid-Atlantic coast on March 5, just as spring tides were peaking, then stalled south of Long Island for three days, killing 34 people and causing an estimated $300 million in damage. Disaster officials estimate damage from a similar storm today would be many times greater due to the surge of coastal development over the last four decades and the fact that about 37 million people, 15 percent of the population, now live between Boston and Washington.

Scenes from the 1962 storm linger today in coastal communities that were hit. In Virginia Beach, Va., 340 homes were destroyed or damaged and 1,000 automobiles ruined by waves that reached 40-feet high. At Long Beach, N.J., a Navy destroyer being towed to port for repairs was flung onto the shore after its towing cable broke. Damage to beaches from Massachusetts to Florida was so great the Army Corps of Engineers began sand pumping on a regular basis to rebuild beaches and provide a buffer against future shoreline damage. [2]

While such extensive damage hasn't been duplicated, a similar northeaster brought about the great New England blizzard of 1978, which dumped 36 inches of snow on the region. Last winter, two giant northeasters within a week of one another hit the Mid-Atlantic coast, eating away large portions of erosion-plagued Assateague Island, Md. and the nearby beach at Ocean City.

Weather patterns spare the West Coast from the storms, though last winter's El Niño-inspired weather brought storm swells that caused comparable damage. Storms resembling northeasters are, however, found off Australia and New Zealand. A southeasterly variant called the *suestado* is often observed off the east coast of South America. Hawaii also experiences storms with strong southerly winds, called *kona*, or leeward, storms.

[1] See H. Michael Mogil, "Nor'easters: Ill Winds of the Atlantic Seaboard," *The Washington Post*, Feb. 12, 1997, p. H1.

[2] See Mary Reid Barrow, "Last Week? No Comparison, Say Residents Who Lived Through the Terrifying Floods of March 1962," *The Virginian-Pilot* (Norfolk, Va.), Feb. 8, 1998, p. A1.

The Corps of Engineers built a series of groins (sea walls built perpendicular to the beach) in the 1960s to halt erosion near the exclusive resort community. The structures cut off the natural flow of sand westward, starving some beaches. When a giant winter storm hit in 1992, the ocean surged through the narrow, starved barrier, destroying 190 of 246 homes in the town. Property owners sued and won a state permit to rebuild the lost homes.

The corps spent $32 million on an emergency project to fill and widen the damaged beach. But then the U.S. Fish and Wildlife Service determined the area was a known habitat of the piping plover, an endangered shore bird, and ruled any construction may violate the Endangered Species Act. The dispute remains tied up in courts. [11]

No single governmental agency is responsible for coastal development. The corps has perhaps the greatest responsibility, charged with keeping waterways navigable and protecting vulnerable shorelines. But states ultimately approve most of the construction on coastlines—usually with the caveat that property owners are building homes and businesses at their own risk.

"The general situation seems to be that everyone should be able to use their property as they wish, as long as they build to established standards," says Rutherford Platt, professor of geography and planning law at the University of Massachusetts, Amherst. "That seems to have gotten away from the original intent of applying land-use planning to steer development away from recognized hazards."

The Coastal Zone Management Act of 1972 was supposed to help states coordinate their policies based on federal guidelines. To date, 32 out of 35 coastal states have approved programs to protect coastal resources. Conservation groups give the act mixed reviews, saying it has encouraged more responsible development away from the water's edge. However, they note the program is voluntary and doesn't require states to consider factors such as how new coastal developments affect water quality.

Some states have enacted tough restrictions, banning certain types of erosion control. North Carolina, South Carolina, Maine, Rhode Island, Texas and Oregon prohibit construction of new sea walls, jetties and other "hard structures" on beaches, citing evidence that they disrupt sand flow. The states continue to support beach replenishment where necessary.

Local officials complain it's difficult to formulate a beach policy because federal agencies with responsibility for coastal management often don't talk to each other. Tony Pratt, beach manager for the Delaware Natural Resources and Environmental Control Department, notes FEMA doesn't get involved when the corps plans an erosion-control project, even though the agency will have to step in if the beach gets stripped away by tides. "It seems to me that it's time that two sister agencies like this [work together]," he says.

Duke geologist Pilkey notes the problem is exacerbated by conflicting philosophies about coastal development. NOAA, through the Coastal Zone Management Act, advocates building away from eroding shores. Similarly, the National Park Service (NPS) has adopted a broad ban on any coastal engineering at its national seashores. But Pilkey says the corps continues to recommend sea walls, groins and jetties be built to fight the sea.

Congress has made several attempts to make sense of the bureaucratic tangles. The latest is legislation sponsored by Reps. H. James Saxton, R-N.J., Don Young, R-Alaska, and Sam Farr, D-Calif., that would create a commission appointed by the White House and Congress to draft a federal ocean policy. "We have a regulatory process dealing with marine environments that generally doesn't work well," says Saxton, chairman of the House Resources Subcommittee on Fisheries Conservation, Wildlife and Oceans.

But Saxton and others acknowledge that any changes the commission recommends will be difficult to implement because of overlapping congressional jurisdictions. In the House alone, flood insurance comes under the purview of the Banking Committee; coastal zone management is left to the Resources Committee while water quality is handled by Transportation and Infrastructure.

"There's little concerted effort anticipating long-term coastal problems," says the National Wildlife Federation's Conrad. "Getting rational decision-making will be difficult because decisions are mostly based on politics, not long-term consequences.

Background

Vacation Boom

America's rush to the shore may have begun in 1802, when promoters from Cape May, N.J., took out an advertisement in the *Philadelphia Aurora*, extolling the beauty of their oceanfront and the spacious lodging available in the town.

The accommodations actually consisted of a large room with a curtain down the center separating men and women. But it was enough to start a phenomenon. By the middle of the 19th century, well-to-do residents of Philadelphia and New York were escaping the sweltering cities in the summertime to promenade on the boardwalks of Atlantic City and Ocean City, N.J. Several presidents visited Long Branch, N.J., and Winslow Homer depicted parasol-toting Victorian ladies looking down on rows of the resort's small, frame bathhouses in one of his famous oil paintings.

Even then, man was battling the advancing sea, with relocation being the only option. In 1888, the owners of the Brighton Beach Hotel in Coney Island, N.Y., jacked up the 6,000-ton landmark, deposited it on 120 rail cars and had six locomotives slowly pull it 2,000 feet inland from the rapidly eroding shoreline. All of the houses in South Seaside, N.J., were built on wooden runners so they could be moved back with the beach. [12]

By the early 20th century, working-class Americans were placing more value on vacation travel to the shore. Developers responded by building more resorts on narrow barrier islands to accommodate the increasing crowds. The construction of a wooden bridge linking Miami Beach to the mainland in 1918 set off a development boom that replaced much of the island's natural dune system with hotels and vacation homes. Even a 1926 hurricane that hit

To Save a Lighthouse

For 128 years, the lighthouse at Cape Hatteras, N.C., has guided mariners through the dangerous shoals known as the "graveyard of the Atlantic." But if nature has its way, the landmark tower with the black and white barber pole stripes may soon join hundreds of sunken ships off the Outer Banks.

Erosion and rising seas have eaten away at land around the lighthouse, which was built in 1870 on a site that at the time was 1,500 feet from the shoreline. Today, waves lap just 120 feet from its brick and mortar base, and scientists estimate that even a moderate-strength hurricane could trigger winds and tidal surges that could topple the structure.

The National Park Service, which owns the lighthouse, wants to move it 2,900 feet away from its present location to a new spot 1,600 feet from the water. The agency recently awarded a $1.4 million contract to International Chimney Co. of Buffalo, N.Y., to begin planning the move. But local residents worry that moving the tower will cause structural damage and instead are backing an alternative plan to build an 800-foot-long steel sea wall and other fortifications around the present site.

In many ways, the debate is a microcosm of the nationwide debate over whether to continue armoring the coastline or retreat from advancing waters. The deliberations gained additional momentum in June, when the Senate Appropriations Committee, at the urging of Sen. Lauch Faircloth, R-N.C., approved $9.8 million in the fiscal 1999 budget to help pay for the move. Advocates of the sea wall—including Rep. Walter B. Jones Jr., R-N.C., whose district includes Cape Hatteras—are trying to obtain $4 million for their project in the House of Representatives.

The 208-foot lighthouse is the tallest in the United States and attracts 250,000 visitors annually. It is the biggest attraction in Dare County, which competes with Nags Head, 50 miles to the north, for summertime tourist dollars. But the lighthouse's

Department of Interior

The Cape Hatteras, N.C., lighthouse

precarious future has caused headaches for federal and local officials for nearly three decades.

In 1970, the U.S. Navy built three groins in front of the lighthouse to protect monitoring equipment that tracked Soviet submarines. The groins slowed erosion but disrupted sand flow, flattening nearby dunes and forming a bay south of the lighthouse. In 1982, a committee of local residents and business interests headed by environmentalist Hugh Morton raised $500,000 and spent $165,000 of it to buy a synthetic bed of seaweed that traps sand running toward the beach and helps rebuild dunes. Morton now wants to spend the remainder of the money to install more artificial seaweed. [1]

The Army Corps of Engineers in the mid-1980s explored building a massive stone wall around the lighthouse but decided the eroding coast would eventually move out from under the structure, leaving it stranded at sea on its own little island.

The push to move the lighthouse began in earnest in 1988, when a National Academy of Sciences committee recommended relocation of the 2,800-ton tower. The panel estimated the shoreline in front of the lighthouse would retreat 157–407 feet by the year 2018 as sea levels rise up to 6.1 inches. The conclusions were affirmed last year in a study by North Carolina State University.

It wasn't the first time relocation of a lighthouse was considered. Southwest Light on Block Island, R.I., was moved from an eroding cliff in 1994, and the Cape Cod-Highland Light was transported to a safe spot on a nearby bluff in Massachusetts in 1996. Neither structure suffered any damage. But those structures were considerably smaller than the 21-story Hatteras tower. Preliminary plans call for engineers to mount the structure and two attached keepers' houses on a track and move them five feet at a time with hydraulic equipment. The entire process would take about a month.

Local residents fear one wrong move would destroy the landmark and leave its 1.25 million bricks scattered along the beach. The prospect of relocating the structure even a half-mile also stirs strong emotions among local residents, who grew up in the lighthouse's shadow and derive their identity from it.

"You tell them there's another option, and nobody wants to move it. I think we can go out and put in a sea wall," says John Hooper, owner of the Lighthouse View Motel in Buxton, which is immediately next to the present site. [2]

Park Service ranger Richard Schneider says he understands the sentimental attachment but that engineering studies show the move can be pulled off without any damage. He adds that disputes delaying the move could make it difficult to obtain restoration money.

"Our feeling is we have the technology to do this and if we wait, as coastal management becomes a bigger issue, Cape Hatteras may fall to the bottom of the list," he says.

---

[1] See Gerrie Ferris Finger, "Fighting for the Light," *Atlanta Constitution,* June 28, 1998, p. 1D.

[2] Quoted in Brian Hicks, "Embattled Beacons: Hatteras Islanders Fight for Their Light," *The Post and Courier* (Charleston, S.C.), June 28, 1998, p. A1.

AP/Patricia McDonnell

Beach erosion threatens seaside homes in Plymouth, Mass., and all along the Massachusetts coast.

the city and killed 243 people didn't deter further growth.

During the Great Depression, the federal government viewed coastal management as a means of creating large public works projects to prevent beaches from washing away. The Civilian Conservation Corps (CCC) in the 1930s constructed a huge dune system on North Carolina's Outer Banks, stretching 115 miles with nearly 142 million square feet of dune grasses and more than 600 miles of sand fencing.

But continued shorefront building led to inevitable encounters with the sea. Those who could afford vacation cottages or larger estates asked for sea walls, revetments (walls of boulders built on the slope of a dune or on an eroding bluff) and groins to protect their properties. Increasingly, local communities turned to the Army Corps of Engineers.

The corps traces its roots to 1775, when George Washington and the Second Continental Congress established a branch of engineers to build fortifications. The elite unit primarily dealt with military matters until an 1824 Supreme Court decision in *Gibbons v. Ogden* gave the federal government authority over interstate navigation. The corps was assigned to dredge the Mississippi and Ohio rivers of obstacles to navigation. Soon, it also was building jetties to protect harbors.

The corps' mission expanded as a series of coastal-protection laws passed from 1930 to 1956 increasingly made the federal government responsible for controlling coastal erosion and protecting private property from floods.

Activity intensified after World War II, when a burst of federally supported road and bridge construction spurred an exodus to destinations like Atlantic City, Virginia Beach, Va., and Hilton Head, S.C., swelling the coastal population by more than 30 million people between 1950 and 1980.

Along the Atlantic Coast, dozens of walls were constructed to halt crashing waves and keep oceanfront property owners safe. Ironically, the massive structures often cut off public access to the beach and began eroding themselves, requiring more fortifications. The result in locales like Asbury Park, N.J., was an unsightly collection of wood, steel, stone and concrete walls and thin strips of beach that Duke University geologist Pilkey refers to as the "Newjerseyization" of the coastline.

In the 1950s, the corps began embracing the alternative concept of "renourishing" heavily developed beaches, piping in sand from offshore bars to build up the buffer zones that tides had washed away. The process gained popularity after a giant northeaster known as the "Ash Wednesday storm" battered the Atlantic Coast from Massachusetts to Florida in March 1962. Tides washed over barrier islands, created new inlets and destroyed $300 million worth of property and some of the most popular beaches. The corps pumped millions of cubic yards of sand to rebuild the strands, restoring the shore's allure.

"We're proud of the projects we've produced for the nation," says Charles Chesnutt, a coastal engineer with the corps in Washington. "If you look at the [dozens of] major projects and the accuracy we've had in cost projections, the report card is very good."

Action by Congress

Congress traditionally treated beach replenishment and sea wall construction the way it did highway projects, authorizing dozens of taxpayer-supported projects in annual spending bills. The process began to come into question

A storm-damaged sea wall in Sandbridge, Va. The Army Corps of Engineers has backed off from using such "hard structures" because they tend to narrow recreational beaches.

Program for the Study of Developed Shorelines

during the administration of President Jimmy Carter, who had grown leery of large-scale water-development projects as governor of Georgia and wanted more executive branch control over the more than $3 billion of annual federal spending on water development.

Carter implemented water-reform policies so that only those projects that were environmentally sound and economically justifiable would be funded. This irked many members of Congress, who tried to authorize some disputed projects by bundling them in annual spending bills and dared Carter to veto them. [13]

After Carter left office, various sides in the debate struck a compromise of sorts in 1982, when Congress passed the Coastal Barrier Resources Act (COBRA) to discourage construction on 186 undeveloped islands, spits and beaches along the Atlantic and Gulf coasts. The law doesn't prevent landowners from building on their land but shifts the risk to the private sector by barring federal spending for roads, bridges and flood insurance.

The question of who should fund long-term shoreline protection was sorted out in 1986, when Congress passed the Water Resources and Development Act (WRDA) and set cost-sharing requirements for shore-protection projects. The federal government agreed to pay 65 percent of the cost of beach rebuilding, covering both initial sand pumping and, after storms eroded beaches, follow-up rounds of pumping every several years over 50 years. States and local governments typically split the remaining 35 percent.

Rising Sea Levels

As lawmakers tried to establish new ground rules for the coast, scientists were issuing a series of reports warning that sea levels were rising at unprecedented levels because of global warming, further threatening new and existing coastal development.

Scientists say sea-level fluctuations are part of a natural cycle, triggered by events like the melting of the great glaciers, which began 20,000 years ago. But many believe the phenomenon has intensified since the Industrial Revolution due to the consumption of more fossil fuels, which release carbon dioxide as a byproduct. The gas traps excess heat that would otherwise radiate into space. As a result, the atmosphere is growing warmer, melting polar ice caps, increasing the volume of the oceans and speeding up erosion. [14]

Geologists blame rising sea levels for the unique tendency of barrier islands to migrate. As storms and normal tidal action pummel dunes on the islands' ocean beaches, sand is pushed to the backside of the islands, or even the lagoons or bays behind them. Over time, the backside becomes the beach and the islands literally fold over on themselves, slowly retreating up the coastal plain.

The action explains how some barrier islands in the Gulf of Mexico have retreated 80 miles over the last century and a half. It also explains how fragile, Mid-Atlantic sites like Assateague Island, Md., are in danger of being "overwashed" in a major storm, making mainland communities more vulnerable to tidal surges. [15]

Federal officials acknowledge that rising sea levels eventually are likely to force a rethinking of coastal development rules. Many coastal structures are designed with the worst flooding theoretically possible over a 100-year period as their basis. But higher sea levels mean a 50-year flood may become as severe as a 100-year flood. Furthermore, FEMA estimates that with a sea-level rise of one meter, the number of households in the coastal floodplain will rise from the present 2.7 million to 6.6 million by 2100.

In such a dynamic landscape, defining what is responsible development can be challenging. A pair of U.S. Supreme Court rulings in the 1990s tried to strike a balance between regulation and property rights by asserting that shoreline property owners still have rights to build in certain areas, even if states try to place them off-limits.

In 1992, the court ruled in *Lucas v. South Carolina Coastal Council* that South Carolina officials violated the takings clause of the Constitution by forbidding developer David Lucas from building on two beachfront lots in an "inlet erosion zone."

Lucas paid $975,000 for the lots on the Isle of Palms in 1986. Before he could put up houses, the state passed a beach management law aimed at preserving the coastline and preventing "unwise development." The high court, by a 6–3 vote, said state regulations that deny a landowner full value of his land constitute a "taking" under the Fifth Amendment and require compensation. [16]

The Supreme Court broadened the definition of property rights two years later, ruling 5–4 in *Dolan v. City of Tigard* that an Oregon landowner didn't have to turn over a portion of his property for a municipal floodplain. The University of Massachusetts' Platt says the cases continue to influence coastal states and municipalities. he says governments may refuse to regulate coastal land because they fear they will have to make large compensatory payments to a landowner.

Current Situation

Budget Battle

If coastal communities benefit the most from shorefront protection, why shouldn't they pay more to protect their homes and businesses?

That's a question the Clinton administration is posing as Congress takes up a biennual reauthorization of the Water Resources and Development Act. The legislation authorizes the Army Corps of Engineers' dredging, navigation and coastal protection projects. After four years of unsuccessful efforts to eliminate some shoreline subsidies, White House officials are taking a softer approach, trying to alter language in the law that would change the federal-local split of costs for sand-pumping on beaches, the most expensive work.

The Clinton administration says it will continue to pay two-thirds of the cost of the initial pumping but wants to reverse the ratio and have states and localities pay 65 percent of the follow-up work, with the federal government picking up the other 35 percent. The administration argues that the constraints of the 1997 balanced-budget agreement it struck with Congress prevent more expensive, open-ended commitments, such as the beach fill at Ocean City, Md.

"The reality is we have to find ways of making the most with all the resources we have," White House Deputy Budget Director T.J. Glauthier told a recent conference of the American Coastal Coalition in Washington.

The administration says it could support total annual expenditures of $80–$100 million on national shore protection out of a total corps construction budget of $1 billion. But Congress appears headed for another fight over budget levels, counting on political support for local water projects during an election year. The Senate in June earmarked $1.2 billion for corps construction—still 15 percent less than 1998 levels but more generous than the Clinton plan.

"It is significantly better than the program proposed by the administration and will allow [construction] to move

FEMA this year launched "Project Impact," a voluntary effort designed to urge communities to reduce life and property damage by studying local zoning and building codes and consulting existing flood maps. It hopes to sign up one community per state by September.

The federal government plans to keep replenishing the 10 miles of shoreline in Ocean City, Md., over the next 50 years at an estimated cost of $500 million.

Adriel Bettelheim

forward," says Sen. Pete V. Domenici, R-N.M., chairman of the Senate Appropriations Energy and Water Development Subcommittee. [17]

Coastal communities are banding together to cheer on lawmakers like Domenici after years of piecemeal lobbying efforts. The American Coastal Coalition was formed by Washington lobbyist Howard Marlowe in 1995 and now has a membership of 140 local governments and property owners associations. In June, the organization convened a "coastal summit" in Washington to pressure House and Senate members to continue supporting shoreline protection as they crafted 1999 spending bills.

The Army corps isn't taking a position on the battle, simply saying it will do the work Congress authorizes.

Pressure on Congress

The lobbying on coastal issues isn't restricted to states and localities. The Pentagon and business groups are pressuring the Senate to hold hearings and ratify the Law of the Sea treaty, a global pact signed by 125 countries that regulates commerce, navigation and exploration on and beneath oceans. The treaty was rejected by then-President Ronald Reagan in 1982 because it required industrialized nations to share revenues from deep-sea mining with developing nations. It has since been amended to satisfy U.S. concerns. President Clinton signed it in 1994 and sent it to the Senate, which has until Nov. 16 to act. [18]

Clinton told the National Ocean Conference in June that approving the treaty is essential for the United States to be a global leader on coastal issues. The U.S. Navy supports the pact because it guarantees passage of surface warships and submarines through strategic waterways without prior notice. The oil industry likes it because it protects offshore drilling rights and similarly guarantees rights of maritime passage for tankers.

However, Senate Foreign Relations Committee Chairman Jesse Helms, R-N.C., has yet to schedule hearings and is said to have concerns that changes to treaty language didn't go far enough. Fueling those suspicions is the fact that the treaty was negotiated under the auspices of the United Nations, a frequent target of Helms' criticism.

On a separate front, environmentalists, developers and their allies are battling over efforts to exclude certain parcels of land from protected status under the Coastal Barrier Resources Act. The 1982 legislation now forbids federal subsidies for coastal protection and post-storm disaster relief on approximately 1.2 million acres of coastal tracts on the Atlantic and Gulf coasts and the Great Lakes.

The dispute was triggered last year when Rep. Tillie Fowler, R-Fla., included language in omnibus parks legislation to lift a ban on federal flood insurance for eight flood-prone areas of Florida, including beachfront property at New Smyrna Beach that was targeted for high-rise condominiums.

Mystery of the 'Cell From Hell'

The small fish began washing up by the thousands on the banks of Chesapeake Bay tributaries last summer bearing distinct red sores and lesions. Swimmers and watermen who came in contact with them soon reported nausea, headaches and, in some cases, memory loss. Within weeks, Maryland authorities declared a health advisory and closed three affected waterways. State and federal scientists began looking for a cause.

The culprit turned out to be the so-called "cell from hell," *Pfiesteria piscicida*, a microorganism that feeds off excessive amounts of nitrogen and phosphorus entering rivers in fertilizers, including manure. The bizarre microbe appears in up to two-dozen biological forms, at least three of them toxic, and symbolizes a new kind of pollution plaguing coastal waterways.

For decades, water pollution was something people could easily see or smell. In 1987, numerous East Coast beaches were closed after raw sewage, hypodermic needles and toxic waste washed ashore. But scientists say those obvious forms of pollution are perhaps less of a threat than increasing amounts of runoff from farms and cities that can foul the wide, shallow estuaries where rivers meet oceans. The new phenomenon is termed non-point-source pollution because contaminants usually end up long distances from their original source.

The U.S. Environmental Protection Agency (EPA) estimates that 32 billion gallons a day of runoff from farms, factories and urban streets and parking lots send nutrients into the water, creating algae blooms that suck oxygen from water and suffocate fish. The nutrients also feed periodic outbreaks of *Pfiesteria*, which was discovered after a series of massive fishkills in North Carolina's Neuse River in the 1980s.

Runoff also contaminates sediment in brackish waterways. A three-volume EPA report released to Congress in January estimated that the sediment in 7 percent of U.S. watersheds, including bays and estuaries, is so seriously tainted that eating fish from those waters would threaten human and animal health. Sites of greatest contamination were clustered around big cities and low-lying regions affected by discharges from farms. [1]

The findings are putting new pressure on agriculture. Farmers spreading fertilizer in Iowa and other Midwestern states are being blamed for disrupting the fishing industry in the Gulf of Mexico, a thousand miles away. Runoff washes down the Mississippi River and creates dead zones, killing slow-moving creatures like crabs while sending more mobile fish and shrimp off for cleaner waters hundreds of miles away. [2]

Some states are tightening regulations. In the wake of last summer's *Pfiesteria* scare, Maryland in

The Coast Alliance, a Washington-based environmental group, and several other conservation organizations blocked the changes by filing a lawsuit, contending the move was a backdoor attempt by developers to grab more land for high-density development. But officials hardly think the chapter is closed. "They obviously want to free up some of the more attractive parcels and use arguments like there was a policy mistake or technical mistake in the way the law was applied," says Jacqueline Savitz, executive director of the Coast Alliance.

Fowler says developers already had installed roads, sewers and water lines on the New Smyrna property by the time the law was passed in 1982 and insists the exemption wouldn't open up more pristine coastal property for development. She adds the exemptions she sought covered land that only totaled 36.4 acres. [19]

Outlook

Ending the Cycle

In Louisiana, officials spend as much as $40 million a year trying to replenish some of the 35 square miles of coastal wetlands that are lost each year to erosion and rising seas. However, figuring out precisely where to shore up the

January unveiled a multimillion-dollar plan to control nutrient runoff from agriculture, including large poultry-processing operations on the state's Eastern Shore. The state was particularly stung because the microbe wasn't previously known to exist in the Chesapeake Bay watershed. Publicity surrounding the outbreak sparked consumer reluctance to buy fish, resulting in $40 million in lost sales to Maryland's seafood industry. Ironically, the outbreak was largely confined to menhaden, an oily, herringlike fish that travels in very large schools but is not typically consumed by humans.

North Carolina for the first time this summer will close waterways hit by fish kills. Virginia state environmental workers this spring began a $7.7. million water-monitoring program to spot signs of *Pfiesteria* and other toxic microbes.

Pfisteria is showing signs of reappearing this summer. An estimated half-million menhaden had died along the Neuse River as of early August, and Maryland officials reported finding lesions on a small number of fish in Chesapeake tributaries.

The Clinton administration has responded by proposing a new clean-water initiative; calling for improved monitoring of non-point-source pollutants like nitrogen and phosphorous; and the placement of buffer zones along 2 million miles of streams to trap fertilizer and manure. White House Chief of Staff Erskine Bowles, a North Carolinian, in August arranged a $365,000 grant to officials in his home state to help rapid-response teams respond to the latest fish kill and is trying to speed up the awarding of $221 million in U.S. Department of Agriculture funds to help pay the state's farmers for field buffers and other pollution controls. [3]

Agriculture interests say they will comply with tougher regulations but suspect they're being unfairly singled out. Some note urban runoff does as much harm to coastal ecosystems, clogging basins with partially treated sewage and silt that kills vital organisms.

"The science with *Pfiesteria* is still catching up. You can't make a quick decision on the fate of agriculture based on *Pfisteria*," says James Perdue, chief executive officer of Salisbury, Md.-based Perdue Farms, Inc., the nation's second-largest poultry producer. [4]

Rep. Wayne T. Gilchrest, R-Md., who represents the largely agricultural Eastern Shore, where the outbreak occurred, says the problem is serious and "speeded up by at least two years" the move to control farm runoff.

"I believe [Perdue] understands that this problem is real and that we are going to have to deal with it, with the industry's cooperation or without it," Gilchrest says.

[1] See "The Incidence and Severity of Sediment Contamination in Surface Waters of the United States," U.S. Environmental Protection Agency.

[2] See Perry Beeman, "Engulfed by Deadly Chemicals Farm Runoff Blamed for Pollution off Gulf Coast," *The Des Moines Register,* May 24, 1998, p. 1.

[3] See Joby Warrick, "States Brace for Fish Kills After 'Cell From Hell' Returns in N.C.," *The Washington Post,* Aug. 6, 1998, p. A2.

[4] Quoted in Tom Horton, "Issue of Agricultural Runoff Isn't Going Away, Perdue says," *The Baltimore Sun,* Oct. 20, 1997, p. 1A.

marshy coast is difficult because few accurate maps exist showing where the water's edge was even a decade ago. [20]

Critics say the situation shows that government planners have only vague ideas about how coastal flooding occurs and how extensively it leads to land loss. Climate records that NOAA and FEMA use to calculate hypothetical "storms of the century" and establish 100-year flood plains are often only 50–100 years old. Yet they define the parameters for development in the nation's fastest-growing regions; federal flood-plain maps are consulted for 15 million mortgage transactions a year.

Slowly, some agencies are launching flood mitigation efforts in an attempt to end the disaster-rebuild-disaster cycle the government long has been accused of promoting. FEMA this year launched "Project Impact," a voluntary effort designed to urge communities to reduce life and property damage by studying local zoning and building codes and consulting existing flood maps. It hopes to sign up one community per state by September.

The agency also is responding to a 1994 mandate from Congress to better evaluate coastal erosion hazards. FEMA and the H. John Heinz III Center for Science, Economics and the Environment in Washington are exploring how dynamic coastline changes can be factored into the sale of national flood insurance policies.

FOR MORE INFORMATION

Federal Emergency Management Agency, 500 C St., S.W., Washington, D.C. 20472; (202) 646-4600; www.fema.gov. FEMA assists state and local governments in preparing for and responding to emergencies, including hurricanes and major coastal storms. Oversees the national flood insurance program.

Coast Alliance, 215 Pennsylvania Ave., S.E., Washington, D.C. 20003 (202); 546-9554; www.coastalliance.org. A nonprofit environmental group founded in 1979 to combat development pressure and pollution on American coastlines.

American Littoral Society, Sandy Hook, Highlands, N.J. 07732; (732) 291-0055. Encourages the study and conservation of marine life and habitat, with a special emphasis on coastal zones.

Coastal States Organization, 444 N. Capitol St., N.W., Suite 322, Washington, D.C. 20001; (202) 508-3860. Represents governors of U.S. coastal states, territories and commonwealths on management of coastal, Great Lakes and marine resources.

American Coastal Coalition, 1667 K St., N.W., Suite 480, Washington, D.C. 20006; (202) 775-1796; www.coastalcoalition.org. Founded in 1996 to fight President Clinton's proposed cuts to beach nourishment programs, this organization lobbies for more federal aid for beachfront communities and property owners.

Longtime observers say such scattered efforts probably won't discourage the continued development at the shore, though they may encourage tougher local building restrictions. But they note the era of balanced budgets and greater concern for the coast is at least making some communities reconsider federal handouts for coastal protection. Avalon, N.J., has spent $8.6 million in local tax dollars to rebuild its five-mile beach after storms. Residents of Dewey Beach, Del., voted to tax themselves to pay for a $1 million beach protection program.

In Massachusetts, Cape Cod voters this November will be asked to endorse a 3 percent property-tax increase to raise money to buy undeveloped land and protect the coast from overbuilding. Similar measures are being proposed in New Jersey, Connecticut, Georgia and the eastern end of Long Island. [21]

"Since World War II, people have marched to the beach, and we've used the greatest technological capabilities we can harness to control nature," says geologist Leatherman. "We're on a collision course. The sea keeps on coming, and properties and buildings are going to be lost." ❖

Thought Questions

1. What restrictions should be placed on coastal development if any?
2. Who should be held financially responsible when natural processes damage coastal developments? What about developments damaged by other natural disasters such as earthquakes, tornadoes, and floods?
3. How might you design a community such that residents and tourists would have easy access to the beach and its recreational opportunities, but that would also be at low risk from coastal hazards?
4. What impacts might coastal development have on wildlife and plant habitats? Should these impacts be considered when developments are up for approval by city, county, and state governments?
5. What factors should be included in a cost-benefit analysis to determine whether shore stabilization activities are cost effective? Over what time period should the cost-benefit analysis be conducted?

Notes

[1] For a more comprehensive overview, see National Oceanic and Atmospheric Administration, "Year of the Ocean: Discussion Papers," a series of background issue papers published by various federal agencies in March 1998 for the National Ocean Conference in Monterey, Calif., U.S. Government Printing Office Document 1998/432–031.

[2] For background, see Rodman D. Griffin, "Threatened Coastlines," *The CQ Researcher*, Feb. 7, 1992, pp. 97–120.

[3] Quoted in Craig Whitlock, "Flooded With Generosity," *Raleigh* (N.C.) *News and Observer*, Nov. 9, 1997, p. A20.

[4] Beth Millemann, "Storm on the Horizon," Coast Alliance and National Wildlife Federation, September 1989.

[5] Op-Ed piece circulated by FEMA to newspapers in July 1997, responding to criticisms of the program by environmental journalist James Bovard.

[6] Quoted in Craig Whitlock, "FEMA to Review its Aid to Risky Coastal Areas," *Raleigh* (N.C.) *News and Observer*, Dec. 10, 1997, p. A1.

[7] See David Bush, Orrin Pilkey and William Neal, *Living by the Rules of the Sea* (1996).

[8] See Tom Horton, "Nourishing Beaches Ad Infinitum," *The Baltimore Sun*, Feb. 13, 1998, p. 2B.

[9] See Laurence Arnold, "White House Tries Again to Halt Sand-Pumping Programs," *The Associated Press*, March 7, 1998.

[10] Quoted in Niraj Warikoo, "Debate Pumped Up Over Sand Fill Plans," *Newsday*, Dec. 12, 1997, p. A57.

[11] See Robert Hanley, "As Beaches Erode, a Debate On Who'll Pay For Repairs," *The New York Times*, April 20, 1998, p. A1.

[12] See Orrin Pilkey and Katherine Dixon, *The Corps and the Shore* (1996).

[13] For background, see *Congress and the Nation*, Vol. V, 1977–1980, Congressional Quarterly (1981), pp. 565–566.

[14] For background see Mary H. Cooper, "Global Warming Update," *The CQ Researcher*, Nov. 1, 1996, pp. 961–984.

[15] For background, see Stephen Leatherman, *Barrier Island Handbook* (1988).

[16] See Kenneth Jost, "Property Rights," *The CQ Researcher*, June 16, 1995, pp. 513–536.

[17] See Allan Freedman, "Defying Administration, Panel Boosts Water Spending, Criticizes Nuclear Regulation," *CQ Weekly*, June 6, 1998, p. 1534.

[18] See Thomas Lippman, "For Sea Treaty, It's Helms or High Water," *The Washington Post*, July 13, 1998, p. A4.

[19] See Krys Fluker, "Lawsuit Brought Against Amendment That Excludes Coastal Development from Environmental Legislation," *News-Journal* (Daytona Beach, Fla.), June 17, 1997.

[20] See Mike Dunne, "Disappearing Louisiana: Computers Aid State in Battle Against its Receding Coastline," *The Advocate* (Baton Rouge, La.), March 8, 1998, p. 1B.

[21] Fred Bayles, "Cape Cod Fighting for Its Soul," *USA Today*, Aug. 18, 1998, p. A6.

Chronology

1900s-1950s

Coastal communities grow with few land-use regulations.

Sept. 8, 1900

A hurricane ravages Galveston, Texas, killing 6,000 people in the worst natural disaster in U.S. history in terms of lives lost. The city responds by building a 3-mile-long sea wall.

1922

New York's Coney Island launches the first recorded beach-replenishment program in the United States in order to protect its well-known amusement park.

1930

Congress establishes the Beach Erosion Board as part of the River and Harbor Act of 1930 to study beach erosion problems at the request of localities. The federal government pays up to half of the cost of studies but doesn't pay any construction costs unless federal property is involved.

1956

Congress expands shoreline protection to authorize federal subsidies for private property if the improvements help protect publicly owned shores, or if there are other public benefits.

1960s-1970s

Government officials become aware of the risks of rapid coastal development and propose measures to regulate growth.

March 5-8, 1962

A northeaster known as the "Ash Wednesday storm" hammers beachfront communities from Massachusetts to Florida, causing an estimated $300 million in damages. The loss of beaches is so severe that the U.S. Army Corps of Engineers is cast in a permanent, new role as the nation's key beach-replenishment agency.

1967

The Stratton Commission leads to creation of the National Oceanic and Atmospheric Administration (NOAA), the ocean research and weather forecasting branch of the U.S. Department of Commerce.

1968

Congress enacts the National Flood Insurance Act to limit increasing expenditures for flood control and disaster relief.

April 1979

North Carolina passes oceanfront setback regulations, setting the standard for other states.

1980s-1990s

As major storms batter the heavily developed East Coast, federal officials try to coordinate programs and deal with land-use and environmental disputes.

1982

Congress passes the Coastal Barrier Resources Act banning federal subsidies for development of 186 undeveloped shoreline tracts on the Atlantic and Gulf coasts.

1986

Congress passes the Water Resources Development Act, mandating that erosion-control projects primarily be used to mitigate storm damage and improve recreational facilities. The federal government eventually agrees to pick up 65 percent of initial construction work.

1987

A National Academy of Sciences report predicts the sea level in the United States will rise at an unprecedented rate, endangering many low-lying coastal areas.

Sept. 22, 1989

Hurricane Hugo rips apart the South Carolina coast, causing 21 deaths and $7 billion in damage.

Aug. 24-26, 1992

Hurricane Andrew slashes across South Florida and Louisiana, killing 14 people and causing more than $20 billion in damage, making it the costliest storm in U.S. history.

June 1994

A White House task force calls for reforming the federal flood insurance program, saying the first priority should be eliminating flood risk.

1995

President Clinton recommends federal participation in termination of new shore-protection projects. Congress rejects the proposal and adds money for new projects.

Sept. 6, 1996

Hurricane Fran hits the Carolina coast, killing 22 people and inflicting $6.5 billion in damage.

June 11-12, 1998

President Clinton and conferees at the National Ocean Conference in Monterey, Calif., discuss developing a national coastal policy. Clinton unveils a series of measures to protect oceans.

At Issue:

Should the federal government continue to subsidize beach-replenishment efforts?

HOWARD MARLOWE—President, American Coastal Coalition

From testimony before House subcommittee on water resources and environment, March 31, 1998.

We believe that federal, state and local investments in beach erosion control and in the proper management of beaches, inlets and shorelines are returned many times over in revenues generated by tourism and commerce, by tax increases inspired by higher property values and incomes, by mitigation of storm wave damage to property and infrastructure and by the elevation of the quality of life for coastal residents and visitors.

Recent studies and surveys have documented the economic value of beaches to specific local communities, regions and states, and while such studies are just beginning to be undertaken on a national level, it is intuitively obvious that thriving local, regional and state coastal economies are necessary factors in a healthy national economy.

We firmly believe that beach nourishment is an effective method of shore protection based on engineering and fiscal criteria. By beach nourishment, we refer to sand placement or sand replenishment. The American Coastal Coalition believes that the federal role in shore protection and beach erosion control is clearly prescribed by current law, including the Shore Protection Act of 1996 (Section 227 of the Water Resources Development Act of 1996). Efforts to substantially reduce and eventually eliminate this role are clearly counterproductive....

Furthermore, we believe the federal government must participate in the management of the nation's sandy shoreline. This includes a strong fiscal commitment to sharing the costs of construction and periodic maintenance of beach nourishment projects with states and/or local governments....

Not every sandy beach is an appropriate candidate for beach nourishment. For the large number which are, there must be an understanding and acceptance of the fact that beach nourishment has as its objective the reconstruction of a beach so that the net loss of sand caused by wave action and storms—and in many cases exacerbated by the existence of inlets and other forms of human intervention—is slowed to a minimum....

Withdrawing from our coastlines is an unacceptable alternative to beach nourishment. The history of mankind is replete with evidence that people are drawn to coastlines for both economic and recreational reasons. Unless the coasts are cordoned off with barbed wire, that attraction will continue.

YES

BETH MILLEMANN—Former executive director, Coast Alliance

From And Two If By Sea, *1986.*

Few areas are less suited to heavy development than the beaches, dunes and islands of the coastal zone, and accordingly in few areas is development more costly in both national and natural resource dollars. Particularly vulnerable to erosion and susceptible to routine flooding and storm damage, the nation's coasts are receding by the foot and yard every year....

Since 1938, more than two dozen hurricanes have cost state and federal governments and taxpayers between $50 million to $2.3 billion in damage per hurricane by flattening homes and businesses on the vulnerable Gulf and Atlantic coasts. In addition, coastal erosion, exacerbated by profligate beach and shore development on all four U.S. coasts, has taken a huge toll on private, state and federal coffers. Add to this the cost of lost coastal wildlife and fisheries from habitat destruction and development, and the total price tag of beachfront homes and those "weekend getaways" becomes very substantial....

The fact that beaches "are not stable entities, but rather, are dynamic landforms constantly subjected to forces that promote erosion and/or deposition" is often overlooked by Realtors and homeowners. Instead, they erect "defense structures" like groins and jetties....

As "protective" structures, these "hard" stabilizing devices generally "benefit only a few and seriously degrade or destroy the natural beach and the value it holds for the majority." Because of the accelerated erosion often caused by hard stabilizers ... some coastal geologists consider them to be most harmful to the beach.

Renourishment, while somewhat of an improvement over hard stabilizers, is "temporary and too costly a solution except for selected communities." Furthermore, "replenishment is often used as an excuse to intensify development," perpetuating the cycle of shore degradation and destruction.

What is often disregarded is the value of the beach itself. Although it "appears to be sterile and devoid of significant life," a large number of plant and animal species depend upon the beach. Dunes abutting the beach shelter a diverse range of life and the upland area as well, acting as "storage areas for sand to replace that eroded by waves" and storms. In turn, the beach supplies sand to renourish dunes.

In this area of constant flux, intensive development is an invitation to disaster—an invitation, unfortunately, too often accepted.

NO

Bibliography

Selected Sources Used

Books

Pilkey, Orrin, and Katharine Dixon, *The Corps and the Shore*, Island Press, 1996.
Two Duke University coastal geologists offer a critical history of the U.S. Army Corps of Engineers and shoreline development in the United States and advocate a strategy of retreat from the advancing ocean.

Millemann, Beth, *And Two If By Sea*, Coast Alliance Inc., 1988.
A guide to coastal management, including numerous facts and figures on coastal pollution and the dangers of overdevelopment. The book highlights states with model legislation in specific areas, such as ocean dumping.

Lencek, Lena, and Gideon Bosker, *The Beach: The History of Paradise on Earth*, Viking Penguin, 1998.
Two scholars of popular culture explore how the allure of the ocean transformed 19th-century American society and grew into a commercial and recreational phenomenon.

Safina, Carl, *Song for the Blue Ocean*, Henry Holt, 1997.
A biologist-writer takes readers on a global tour of the world's oceans, arguing that the unregulated global economy is placing enormous pressures on the sea.

Hinrichsen, Don, *Coastal Waters of the World: Trends, Threats and Strategies*, Island Press, 1998.
A journalist-environmental consultant surveys 13 coastal areas around the world and projects future conflicts between development and nature.

Articles

Nash, Betty Joyce, "Shrinking Beaches, Swelling Problems," *Cross Sections* (quarterly of the Federal Reserve Bank of Richmond), Vol. 13, No. 2 (summer 1996).
An overview of questions arising from erosion along the Atlantic coast, including what prevention strategies can work, and who should pay.

Whitlock, Craig, "Flooded With Generosity," *Raleigh* (N.C.) *News and Observer*, Nov. 9, 1997.
Lavish amounts of federal disaster aid helped rebuild North Carolina's coast after Hurricane Fran smashed into the state in 1996. But critics say the subsidies were a waste of money and send the wrong message to developers.

Garland, Greg, "Down The Drain," *The Advocate* (Baton Rouge, La.), Oct. 19–21, 1997.
A three-part series examines how the federal flood insurance program spends more than homes are worth to repair flood damage.

Carr, Edward, "The Sea: A Second Fall," *The Economist*, May 23–29, 1998.
A survey article argues that while the sea once seemed infinite in its bounty, it now is suffering from overfishing and pollution and needs care and maintenance.

Haggerty, Maryann, "A Gathering Storm Over Assateague: The Forces of Beach Erosion, Tourist Trade at Odds on Island," *The Washington Post*, May 9, 1998.
Two winter storms that battered Assateague Island, Md., last winter raise questions about how to deal with the relationship of man and nature at the edge of the sea.

Weber, Peter, "It Comes Down to the Coasts," *World Watch*, Vol. 7, No. 2 (March/April 1994).
A Worldwatch Institute research associate argues that society will have to begin altering patterns of settlement and development to avoid overstressing coastlines.

Reports

Houston, James, "Beachfill Performance," *Shore and Beach*, July 1991, pp. 15–24.
A U.S. Army Corps of Engineers coastal engineer outlines reasons periodic beach renourishment is a successful way to protect shores with few accompanying environmental problems.

Pilkey, Orrin, "The Engineering of Sand," *Journal of Geological Education*, 1989, Vol. 37, pp. 308–311.
Duke University geologist Pilkey presents a study of East Coast barrier islands demonstrating that parameters used to design beach-replenishment projects don't work and that predictions of durability are incorrect.

Leatherman, Stephen, "Barrier Island Handbook," University of Maryland Coastal Publications Series, 1988.
Geologist Leatherman presents a thorough overview of how barrier islands behave in response to development and natural erosion, as well as recreational impacts and development potential.

The Next Step

Additional information from UMI's Newspaper and Periodical Abstracts™ database

Beach Replenishment

"Sand Subsidies Costly, Futile," *USA Today*, June 2, 1998, p. A12.
Waterfront communities across the nation increasingly are looking to Washington to shore up disappearing beaches. Since 1965, the federal government has pumped out more than $1 billion to replenish more than 1,300 eroding beaches—often to see them wash back into the sea. Local communities and states typically chip in only 35 percent of the cost of any project.

"Sandy Hook Seeks Pipeline To Help Fight Beach Erosion," *The New York Times*, Jan. 3, 1998, p. B5.
The federal government has been asked to approve a proposal for the first major replenishment of Sandy Hook's beaches since 1989. Securing federal funds to combat beach erosion is a tough sell, because the Gateway National Recreation Area at Sandy Hook is only one of 365 units in the federal system. The cost of replenishment can range from $12 million to $20 million, a major chunk of the Department of the Interior's $60 million annual budget for the National Park Service.

Hanley, Robert, "As Beaches Erode, a Debate On Who'll Pay For Repairs," *The New York Times*, April 20, 1998, p. A1.
The federal government, which shorefront residents have always considered their savior, now is balking at playing a major financial role in the restoration of beaches. For the federal fiscal year starting Oct. 1, the Office of Management and Budget has earmarked $3.7 million for the Army Corps of Engineers to rebuild beaches in New Jersey and on Long Island. Coastal business owners and local officials and their allies on Capitol Hill want $50.6 million to continue existing beach projects and to study the need for new ones.

Luoma, Jon R., "Oceanfront Battlefront," *Audubon*, July 1998, pp. 50–56.
Extraordinary measures are being taken to keep beaches and coastal towns from being swallowed up by the sea. The author examines how far people should go to protect the land and who should foot the bill for the work.

McLaughlin, Jeff, "Coastline Ebb, Flow Mapped in Study," *The Boston Globe*, June 15, 1997, p. W1.
Massachusetts' Coastal Zone Management office published a landmark study in June 1997 of the changes that have occurred over the past 150 years along the state's tidal shorelines.

Flood Insurance

"A Bargain in Flood Insurance; Federal Discount Could Benefit Many Residents of Region," *Los Angeles Times*, Nov. 17, 1997, p. B4.
It's well-known that certain Southern California areas are at higher risk of flooding. In Los Angeles County, for example, parts of Long Beach, Lynwood and Montebello have been on the federal government's special flood hazard area maps since the early 1980s. In Orange County, the vital but unfinished Santa Ana River Flood Control Project is designed to protect what has long been considered one of the most vulnerable flood plains west of the Mississippi River.

Edwards, Brian, and Don Hunt, "High and Dry; The Right Flood Insurance Can Help You Keep Your Head Above Water," *Chicago Tribune*, May 8, 1998, p. 1.
Available through most home or auto insurance agents, flood insurance is underwritten by the National Flood Insurance Program, a part of the Federal Emergency Management Agency. Traditional homeowners' and renters' insurance will not cover damage caused by overland flooding—a fact that is learned too late by hundreds of people each year. By federal law, homes in designated flood plains must have federal flood insurance as part of their mortgage loan. A flood plain is any area that has experienced flooding in the past, or any area that may not have flooded before but could in the future due to certain changes in the land structure, such as new property development.

Hilmes, Marsha, "Girding for Floods," *The Denver Post*, Sept. 7, 1997, p. F1.
Communities across the country have been participating in mitigation activities for years to try to prevent damages from flooding and other natural disasters. Mitigation entails actions or activities designed to help protect the citizens of a community. For example, since 1990 Fort Collins has had an education and outreach program focusing on flooding. Activities associated with this program include: Flood Awareness Week with informational displays at the Public Library and City Hall, sending special

mailers to all floodplain residents and to members of the Board of Realtors and publishing flood-related articles in a newsletter that goes to all utility customers.

Tharpe, Gene, "Flood Defense Needs More Than Umbrella," *Atlanta Journal-Constitution*, March 29, 1998, p. R8.
When creeks and rivers start rising and cause millions of dollars in flood damage, as they did this month in Georgia, some homeowners start wondering about buying flood insurance to protect their homes. Those who rely on their regular homeowners insurance policy to cover flood damages normally will be disappointed. "A homeowners insurance policy for traditional site-built homes does not provide coverage for flood damage, although it does cover several other types of water damage," said David Colmans of the Georgia Insurance Information Institute.

Government Involvement

"U.S. Coastline Calamities Under Scrutiny," *New Orleans Times-Picayune*, Feb. 14, 1998, p. A2.
The National Oceanic and Atmospheric Administration, troubled by declining fisheries and rampant coastal development, is beginning an effort to identify the problems that afflict America's shoreline and marine ecosystems. "There is an urgent need to nail down the causes and extent of the problems that plague our coastal areas so solutions can be found," NOAA Administrator D. James Baker said Friday in announcing plans for the "State of the Coast" study.

Barnum, Alex, "Legislature to Consider Flood of Coastal Protection Bills; Effort Called Strongest in 25 Years," *San Francisco Chronicle*, March 24, 1997, p. A17.
A bipartisan coalition of coastal state legislators has introduced a wave of bills that is the most ambitious effort to reform California's coastal and marine laws since voters created the Coastal Commission in 1972. That commission regulates development and ensures public access to the coastline. The package includes bills that would strip regulatory control over fishing from the Department of Fish and Game and give it to a newly created commission, create a comprehensive monitoring program of coastal water pollution and establish marine sanctuaries to protect marine life.

Leitner, Peter M., "A Bad Treaty Returns: The Case of the Law of the Sea Treaty," *World Affairs*, winter 1998, pp. 134–150.
In 1982, President Reagan announced that the United States would not become a signatory to the United Nations Convention on the Law of the Sea (UNC-LOS). The author discusses how U.S. participation in the UNC-LOS was a giant step forward in the continuing delegation of U.S. foreign policy to the U.N.

Sea Walls

"Our Sandy Sentinels," *New Orleans Times-Picayune*, April 16, 1998, p. B6.
There are many avenues to trying to counter Louisiana's serious problem of coastal erosion and wetlands loss, but perhaps the most direct method is to improve and maintain the "sea wall." The sea wall in this instance is not man-made, but nature's line of barrier islands. A new federal-state project will rebuild the losses over the rest of this year under the Breaux Act, named for Sen. John Breaux, D-La., which provides money for anti-erosion projects. With the feds paying 85 percent and the state 15 percent of the $28.5 million cost, the project will pump more than 11 million cubic yards of sand from Terrebonne Bay to bolster three of the islands: Trinity, Whiskey and East.

Reid, Alexander, "As Waves Lash, the Seawalls Crumble; Towns Tossed by Rising Costs of Repairs," *The Boston Globe*, Feb. 15, 1998, p. W1.
The wave-battered sea wall along Oceanside Drive, near Sand Hills in Scituate, collapsed last month, opening a breach wide enough to drive several trucks through. Oceanside Drive's sea wall is one of several decrepit barriers along Scituate's 20 miles of coast, and all are a serious worry, said the public works director, Anthony Antoniello. In Hull, where the town is trying to pressure the state to fund repairs to the Point Allerton sea wall, Town Manager Philip Lemnios said, "Most of these walls are of the same vintage, and they're falling apart at the same time. They're at the end of their useful lives."

Storms

Argetsinger, Amy, "After Another Ocean Storm, a New Wave of Expense," *The Washington Post*, Feb. 22, 1998, p. B1.
Even in a stormless season, coastal officials spend millions in public funds annually trying to hang onto the beachfront that the ocean washes away. By the mid-1980s, years of erosion had sent the waters of the Atlantic rushing within 20 feet of Ocean City, Md., homes and hotels at high tide. In Bethany Beach, Del., two hurricanes left a gray, pitted beach so narrow that tourists had to squeeze between boardwalk pilings to find a sandy spot to sit. The most ambitious fortification project was undertaken in Ocean City. In 1988 engineers spent $14 million in state and local funds pumping 2.5 million cubic yards of sand from the ocean bottom to the beach.

Mogil, H. Michael, "Nor'easters; Ill Winds of the Atlantic Seaboard," *The Washington Post*, Feb. 12, 1997, p. H1.
The last week of December brought record-breaking winter storms to the Pacific Northwest. They are Nor'easters, known for their strong winds that blow from the northeast. As these winds blow ashore from the North Atlantic, they often bring high tides and violent waves, causing significant coastal flooding and beach erosion. Just ask the folks in Ocean City, Md. Almost every year there, nor'easters keep the Army Corps of Engineers and local officials busy dredging sand from offshore to replenish the beach.

Vigue, Doreen Iudica, "With Seas Cresting 16 Feet Above Normal, Breakers, Sand Dunes, and Seawalls Were No Match for the Swollen, Raging Atlantic," *The Boston Globe*, Feb. 6, 1998, p. B4.
For residents of Revere, Winthrop, Scituate and Hull, the record-breaking snowfall from the Blizzard of '78 was almost an afterthought. The wind and the waves were their greatest tormentors, combining to decimate 340 houses, damage 6,000 others and take the lives of the five-man crew of a Gloucester pilot boat, a 5-year-old girl and a 62-year-old man.

Over the last three decades the United States has become increasingly dependent on imports of foreign oil (for a discussions on origin and use of petroleum, see Merritts, De Wet, & Menken, pp. 331–338; Press & Siever, pp. 510–513). Imports constituted 36% of oil used in the United States in 1973, and grew to 56% by 1999. The intervening 25 years did not experience a monotonic increase in the amount of oil imported, however. In fact, after peaking at around 48% in 1977, foreign oil imports actually declined throughout the late 1970s and early 1980s. The causes of oil import fluctuations are many, but as Mary Cooper points out in the following article, most have been related to politics. In 1973, the United States supported Israel in the Yom Kippur war against its Arab neighbors. In retaliation, the Organization of Petroleum Exporting Countries (OPEC) levied an embargo against the United States, cutting the United States off from its oil. Long lines formed at gas stations and the price of oil spiked to ten times its pre-embargo level. The Islamic revolution in Iran in the late 1970s and the ten-year-long Iran-Iraq war caused further problems with the oil supply.

Recognizing the vulnerability of the United States to oil shortages brought on by political turmoil, President Jimmy Carter signed the National Energy Conservation Policy Act into law in 1978. This legislation supplied funds for research into alternative energy resources and for promotion of conservation measures (see Merritts, De Wet, & Menken, pp. 357–358; Press & Siever, pp. 512–513, 521–522). Increased domestic oil exploration, coupled with development of nuclear power plants and some renewable energy resources such as solar and wind power, led the United States to decrease its dependence on foreign oil imports until the middle 1980s. Conservation also played a large role as Congress legislated that new cars must achieve fuel efficiencies of 27.5 miles per gallon by 1985, more than double the previous average of 13 miles per gallon.

Little progress has been made toward decreasing fuel consumption since the mid-1980s. The auto fuel efficiency standards legislated by Congress do not apply to light trucks, a fact that has allowed the auto industry to develop sport utility vehicles—gas guzzling cars that in 2000 accounted for more than 50% of new cars sold in the United States. Furthermore, a financial crisis in Asia led to a decrease in industrial production, creating a glut of oil on the world market in the late 1990s. U.S. consumers enjoyed gas prices lower than any seen in decades when adjusted for inflation. The cheap price of energy led to an increased per capita energy consumption in the United States and to increased importation of foreign oil. Whereas U.S. consumers benefited from the price reductions, OPEC nations saw their profits disappear. Many domestic oil companies were likewise affected, and when the price of a barrel of oil fell to less than half of what it cost to produce, many companies were forced to cap wells and lay off workers. Recently oil prices have again soared as OPEC nations have tried to recoup their losses by driving oil prices upward through decreases in production. Thus Americans are again feeling the pinch from our petroleum-dependent economy. Climate change scientists have also sounded the alarm that ever-increasing amounts of carbon dioxide produced by fossil fuel combustion threaten to irrevocably alter Earth's climate, bringing higher sea levels from melting ice sheets, changed precipitation patterns, and the possibility of more devastating hurricanes (see Merritts, De Wet, & Menken, pp. 401–404, for a discussion of global warming).

Initially it appeared that expansion of nuclear power would answer our nation's energy needs, while at the same time minimizing the atmospheric pollution that leads to global warming. Unfortunately, nuclear power production generates radioactive wastes that remain deadly to humans and other lifeforms for hundreds of thousands of years (see Merritts, De Wet, & Menken, pp. 346–350; Press & Siever, pp. 516–517). Faced with the problem of how to isolate and store this waste in a manner that will not cause damage to ecosystems and future human inhabitants of the Earth, nuclear energy has lost popularity, and the search for more environmentally benign forms of energy continues. In coming years solar and wind energy promise to gain importance as will biomass fuels and new technologies such as hydrogen fuel cells (see Merritts, De Wet, & Menken, pp. 350–356; Press & Siever, pp. 517–521).

The Politics of Energy

How should Congress handle U.S. energy problems?

Record low oil prices and deregulation of the electric-utility industry are prompting the Clinton administration and Congress to review the nation's quarter-century-old energy policy. While low prices please consumers, they devastate oil companies, which are seeking tax credits and lower royalty payments for oil pumped on public land. At the same time, environmentalists are seeking greater federal support for the development of solar energy and other renewable sources. They say that reducing fossil-fuel consumption is the best way to reduce global warming. Deregulation of the electric-utility industry is also causing fundamental changes in the U.S. energy picture. This year, the new Congress will take up the longstanding question of what role the federal government should play in deregulation.

Plans by Exxon Corp. to acquire Mobil Corp. come amid the lowest oil prices in 25 years.
Ira Schwarz/Reuters

THE ISSUES

57
- Should the government do more to assist U.S. oil producers during the current oil glut?
- Should Congress mandate electric-utility deregulation by a fixed deadline?
- Should the government do more to encourage the use of renewable energy?

BACKGROUND

63 **Oil Crises Fuel Policy**
Disruptions of oil supplies in the 1970s prompted the country's first sweeping energy-policy initiative.

66 **From Crisis to Glut**
By the mid-1980s, demand for oil had fallen, and oil exports flooded the market.

66 **Reminder From Iraq**
Saddam Hussein's invasion of Kuwait in 1990 underscored U.S. dependence on Middle Eastern oil.

CURRENT SITUATION

67 **Clinton's Energy Policy**
The president's plan calls for reduced greenhouse emissions and deregulation of energy markets.

69 **Utility Deregulation**
Deregulating utilities could save billions per year.

71 **Assessing the Risks**
Some experts worry that deregulation will compromise the reliability of the U.S. electricity grid.

72 **Oil Industry Mergers**
To survive the slump in prices, big oil firms are buying the competition.

OUTLOOK

72 **Overhaul Unlikely?**
Congress appears unwilling to undertake major energy-policy reforms.

SIDEBARS & GRAPHICS

58 **Use of Imported Oil Increased Since 1973**
Rising consumption was a major factor.

59 **Oil Produces Biggest Share of U.S. Energy**
Half the energy used in the U.S. came from petroleum.

63 **Gasoline Prices Dropped**
The decline was about 50 percent in last 20 years.

64 **Nuclear Power vs. Nuclear Waste**
Storing waste poses a major problem for nuclear power.

67 **Coal Produces Most U.S. Electricity**
Nuclear energy is second.

68 **Do Diplomacy and Energy Security Mix?**
U.S. foreign policy and energy needs often conflict.

70 **Global Warming Boosts Renewable Energy**
Less-polluting energy sources sought.

75 **Chronology**
Key events since 1973.

76 **At Issue**
Will oil-company mergers lead to higher fuel prices?

FOR FURTHER RESEARCH

77 **Bibliography**
Selected sources used.

78 **The Next Step**
Additional sources from current periodicals.

Note: For more information on this topic, please see the following pages in Press and Siever's *Understanding Earth,* Third Edition: pp. 510–513, 516–522; and in Merritts, De Wet, and Menking's *Environmental Geology:* pp. 331–338, 346–356, 357–358.

The Politics of Energy

By Mary H. Cooper

The Issues

It's little wonder motorists are smiling these days. A large mocha latte at Starbucks will set you back $3.50 (or about $21 a gallon); fancy spring water goes for $3 a gallon or more. But at gas stations around the country, a gallon of unleaded regular gas is less than a buck.

The low prices were sparked by a global oil glut caused mainly by falling demand in financially depressed Asia and unusually warm weather in Europe and North America. The unprecedented oversupply has knocked down the price of crude oil from $17–$21 a barrel in the 1980s to around $10 today, and rocked the big oil companies.

"Gasoline prices, when corrected for inflation, are not only lower than at any time since during World War II, they're actually lower than the average price during the Great Depression," says Daniel Yergin, chairman of Cambridge Energy Research Associates Inc.

Today's oil glut stands in stark contrast to the situation a quarter-century ago, when a Middle Eastern oil embargo and subsequent quadrupling of oil prices launched a decade of stagflation—high inflation and stagnant output—in the United States. The energy crises of the 1970s prompted President Jimmy Carter to introduce the country's first sweeping energy policy, which largely focused on reducing the country's dependence on foreign oil.

Administered by the newly created Department of Energy, the policy encouraged purchasing crude outside the Middle East, reducing consumption by making cars, buildings and appliances more energy-efficient and developing alternative, especially renewable, energy sources.

The policy has paid off. Diversification of overseas suppliers has loosened the grip that the Organization of Petroleum Exporting Countries (OPEC) once had on world oil supplies. Non-OPEC producers, such as Mexico and Canada, today provide 54 percent of U.S. oil imports, reducing the country's vulnerability to sudden interruptions in oil supplies. And advances in energy efficiency have reduced the energy requirements of myriad products.

But cheap oil carries a hidden price that threatens to unravel the gains brought by the nation's longstanding energy policy. No longer prodded by high oil prices to conserve energy, American motorists are renewing their love affair with gas-guzzling vehicles like the ubiquitous sports utility vehicle. Despite advances in energy efficiency, per capita energy consumption has increased in recent years, actually surpassing the pre-embargo level in 1973. And as domestic oil production continues to decline, U.S. dependence on foreign oil has actually grown, from 36 percent in 1973 to about 56 percent today. That makes the United States more vulnerable than ever—economically and militarily—to disruptions in foreign oil supplies. [1]

"Oil is different from any other energy issue," says David M. Nemtzow, president of the Alliance to Save Energy, a Washington, D.C., coalition that promotes energy efficiency. "You don't go to war over electricity or natural gas. You go to war over oil. Today oil costs 10 bucks a barrel, and that's all people think about. They don't think about the Fifth Fleet."

Low oil prices are also hurting the oil industry. Hit by falling profits, big oil companies such as Exxon Corp. and Mobil Corp. are negotiating mergers and laying off thousands of workers. With some domestic crude oil prices now below $8 a barrel, many domestic producers can no longer afford to keep their low-yield, "marginal" wells pumping. Together, U.S. oil companies are clamoring for government relief.

"It's very important that the federal government continue its long-held policy that oil is a strategic commodity that is absolutely vital to our national security," says Red Caveney, president of the American Petroleum Institute (API), the industry's main membership association.

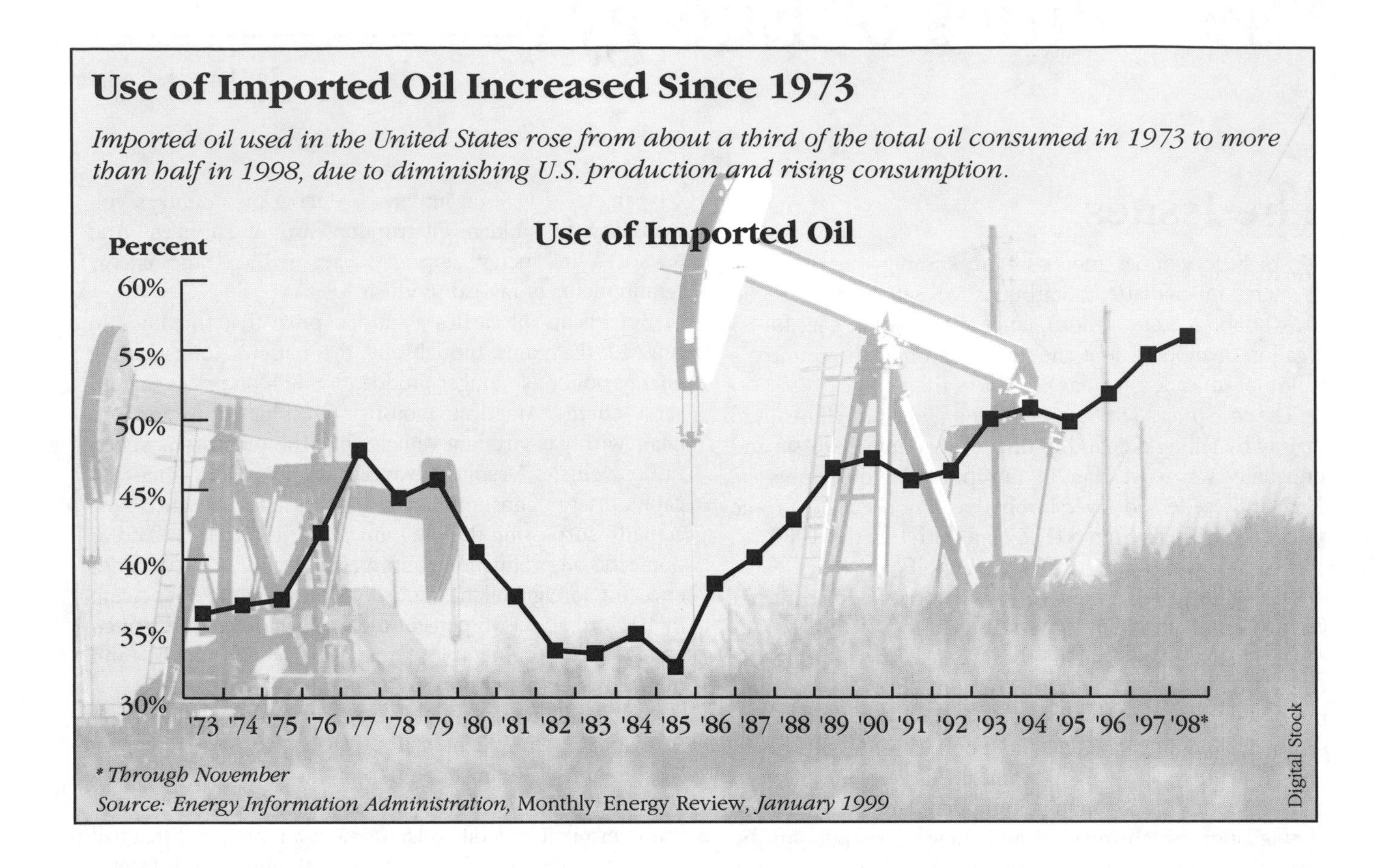

Use of Imported Oil Increased Since 1973

Imported oil used in the United States rose from about a third of the total oil consumed in 1973 to more than half in 1998, due to diminishing U.S. production and rising consumption.

** Through November*
Source: Energy Information Administration, Monthly Energy Review, *January 1999*

"It is vital that the government have policies that protect the U.S. oil and gas resource base."

Falling oil prices are not the only change affecting the U.S. energy sector. Deregulation continues to profoundly change the way consumers purchase energy. The natural gas industry was the first to be deregulated, in the 1980s. Now, it's electric utilities' turn. More than 20 states are now in the process of loosening regulations that for the past 60 years have restricted utilities to the production and distribution of electricity, and set rates for customers.

"Low oil prices, combined with deregulation and competition in the natural gas and power businesses," Yergin says, "constitute the biggest transformation of the energy business since the 1970s."

Still in its infancy, electric-utility restructuring may eventually have an even greater impact on consumers than the court-ordered breakup of AT&T and telecommunications deregulation in the 1980s. "It will have a far greater effect," says Ellen Berman, president of the Consumer Energy Council of America Research Foundation, which studies the social and economic impact of energy policy. "Utilities will be free to provide not just electricity but cable, telephone and Internet access as well. Consumers conceivably will have to choose among hundreds of providers of these services. The question is, 'How do you go about making a choice?' "

Some experts maintain that utility restructuring is undermining the reliability of the U.S. power grid by forcing utilities to place competitiveness for market share above maintenance of surplus power capacity. Last June, for example, storm damage to power lines and a surge in demand during a hot spell caused the price of wholesale electricity in the Midwest to spike and brought the region close to a major power outage. "What happened last summer is going to happen again, and it's going to happen even bigger in the future," says Harry Chernoff, senior economist at Science Applications International Corp., a high-technology research and engineering company in McLean, Va. "The current approach to deregulation has too much economic motivation, and too little engineering and prudence, associated with it."

In addition to falling prices and deregulation, global warming is also causing fundamental changes in the U.S. energy picture and posing difficult challenges to policy-makers. Virtually unheard of 25 years ago and still the subject of controversy, the heating of Earth's atmosphere is widely believed to result principally from the burning of fossil fuels, especially oil and coal. In December 1997, in an effort to slow global warming, President Clinton signed the Kyoto Protocol, which called for reductions in emissions of carbon dioxide and other so-called greenhouse gases. Policies being discussed to meet the treaty's goals include more federal funding of research into solar, wind and other renewable energy sources and measures to convert electric utilities from coal to natural gas and other less-polluting fuels. Not surprisingly, opposition to such policies runs strong in the oil and coal industries, as well as in Congress.

Oil Produces Biggest Share of U.S. Energy

Petroleum products were the source of nearly half the 7.3 quadrillion Btu's of energy consumed in the U.S. in October 1998 (graph at left). The industrial sector accounted for nearly 40 percent of the total energy used (graph at right).

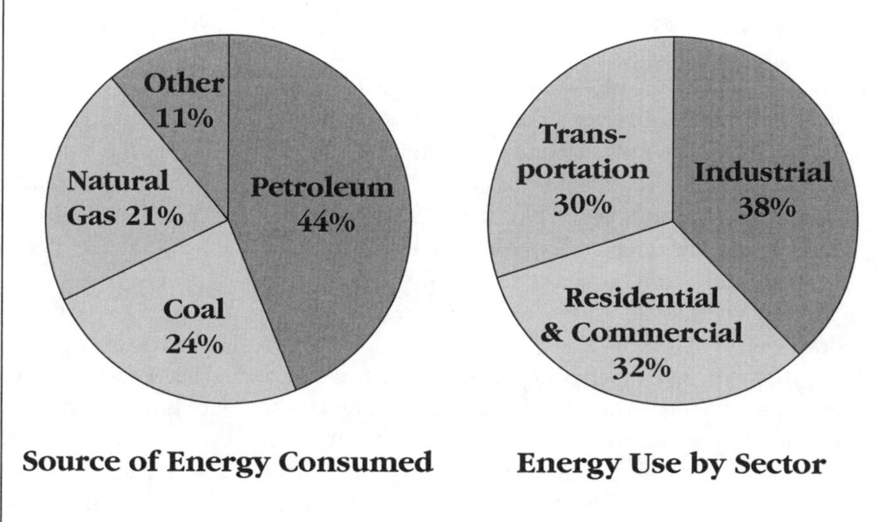

Source: Energy Information Administration, Monthly Energy Review, *January 1999*

"At the same time that the utility industry is changing, so must the regulatory system and the underlying legal structure," says L. Andrew Zausner, a lawyer who specializes in legislative and regulatory affairs at the Washington law firm of Dickstein Shapiro Morin and Oshinsky. "The problem is that we have a whole range of other, interrelated things that our political process is not really good at dealing with, but that all need to be addressed concomitantly. They involve everything from what to do about renewables to global climate change."

In light of the sea changes in the energy sector over the past 25 years, many industry leaders and experts in the field are calling for a comprehensive review of U.S. energy policy. "The question remains, how secure is our energy supply?" then-Rep. Dan Schaefer, R-Colo., former chairman of the House Commerce Subcommittee on Energy and Power, said at a hearing last October. "Do we have a coherent, effective national policy with respect to energy, or are we headed for another crisis? I believe the Congress, the administration and the private sector should work together today to assure that we are doing whatever is necessary to assure that Americans' energy needs will be adequate and secure, not just today, but 10, 25 and 50 years down the road."

As lawmakers and administration officials contemplate changes in energy policy, these are some of the issues they are considering:

Should the government do more to assist U.S. oil producers during the current oil glut?

Low oil prices have had the greatest adverse impact on the nation's "independent" oil producers, which derive most or all of their income from domestic oil production. These companies, such as Philips Petroleum Co. and Conoco Inc., now collect only about $8.50 for a barrel of oil that costs up to $15 to produce, forcing many to close their wells and lay off workers—more than 29,000 employees last year alone. During the first half of 1998, 48,702 wells in 23 states were taken out of production, according to the Interstate Oil and Gas Compact Commission, a 29-state body that monitors domestic production. The commission also reports that the number of rigs drilling for gas and oil in the United States dropped from almost 4,000 in 1981 to fewer than 700 last year. [2]

The beleaguered domestic oil producers are among the most vocal critics of current energy policy, which they say has left them out in the cold. "After the loss of hundreds of thousands of American jobs, billions in education-supportive tax revenues and the spending of tens of billions of defense dollars in the Persian Gulf, we have yet to hear a galvanized political agenda for a national energy policy," Dewey F. Bartlett Jr., vice president of the Oklahoma Independent Petroleum Association, wrote recently in *The Daily Oklahoman*. "Our family dog, Trooper, has more loyalty to the hand that feeds him than our political leaders possess to the industry that literally runs our economic machine." [3]

Domestic producers are calling for a range of policy changes to help them survive, including a tax on imported oil, tax breaks for the industry and a commitment by the federal government to take domestic oil in lieu of royalty payments—the fees companies pay the U.S. government for pumping oil on federal lands—to augment the stocks of the Strategic Petroleum Reserve, set up by Congress in 1975 to ensure the continuity of oil supplies in case of a temporary cutoff in imports.

Energy experts agree that policy changes are needed to help the independents. "Oil is a national asset," Yergin says. "Some kind of measures to enable them to get through this period of very low oil prices would be helpful. There should be some things to give them some breathing space and enable them to survive in a very adverse environment."

The oil industry is divided over what policy changes would be most beneficial, however. Big oil companies, with their extensive operations overseas, oppose a tax on oil imports. But they agree with the rest of the industry on the need for other measures. "Marginal wells produce less than 15 barrels a day, so they're very small," says Caveney of the API. "But people don't understand that, in the aggregate, their output fairly well matches the total imports to the United States from Saudi Arabia." Saudi Arabia is the third-largest supplier of U.S. oil imports after Venezuela and Canada.

The API supports a bill introduced last year and again last month by Sen. Kay Bailey Hutchison, R-Texas, called the United States Energy Economic Growth Act. The measure would provide a $3-a-barrel tax credit for the first three barrels of daily production from an existing marginal well. The credit would be dropped when oil prices rise above a certain amount. A similar credit would apply to producers of natural gas. Other measures supported by the industry include tax relief for producers who reopen inactive wells and assurances that oil and gas producers will not lose their leases when wells are inactive for long periods.

"There isn't a silver bullet that anybody has come up with that is going to address low commodity prices," Caveney says. "So we're looking at a whole series of actions. All of them are admittedly very small, but they would help keep the domestic production base firm and viable for use in the future."

While oil companies argue for ways to increase domestic output and ensure oil supplies, environmentalists and many energy experts favor policy changes aimed at reducing the country's reliance on oil from any source. "Right now, the oil economy is awash in supplies, and the only debate is over when it will run out," says Nemtzow of the Alliance to Save Energy. "Some say five years, some say 30 years. The consensus seems to be at around 10 years, in light of the slowdown in consumption in East Asia."

Even with no changes in current policy, Americans could go a long way toward avoiding future energy shocks by simply reducing energy use, the alliance says. It calculates that the federal government alone wastes $1 billion a year by failing to meet the efficiency improvements mandated by the 1992 Energy Policy Act.

The best way to avoid a future crisis in energy supplies, Nemtzow says, is to improve energy efficiency. "There was a sense in the 1970s, when oil supplies were running out, that we would have to do without," Nemtzow says. "Think of Jimmy Carter sitting in front of a fireplace with a sweater on." Today, technological advances have made it easier to save energy. "We don't ask people to sit in the dark or be cold," he says. "We ask them to use fluorescent lighting, which is much more efficient. Efficiency means doing the same or more with less."

To help wean the country from its dependence on fossil fuels, the alliance calls for more federal funding of research into new energy-efficient technologies, tax breaks for individuals, businesses and utilities investing in those technologies and improved information and labeling requirements to help them do so.

Should Congress set a fixed deadline for electric-utility deregulation?

California and about 20 other states have taken steps to loosen decades-old rules that have protected electric utilities as state-regulated monopolies providing exclusive service to defined geographical areas. In exchange for protection from competition, the industry was required to ensure the reliability of the electricity grid and accept rate structures established by state public utility commissions. But over the past decade, technological advances have enabled other power generators to provide electricity at competitive costs, eroding the rationale for regulation. By 1998, nearly two-thirds of the more than 3,000 electric utilities in the nation had ceased generating electricity altogether and were buying all their power from non-utility generators and other utilities. [4]

"I doubt there is one person in this room who believes that retail electric competition is not inevitable," Energy Secretary Bill Richardson told utility executives at a meeting

When Iraqi forces occupied Kuwait on Aug. 2, 1990, the United Nations imposed an embargo on Iraqi oil exports, which pushed oil prices from $13 a barrel to $40. Above, the Iraqi-Turkish pipeline.

Faleh Kheiber/Reuters

of the Edison Electric Institute last year. "While there may be bumps along the way, the industry is clearly headed down a new and exciting path." [5]

The question is what role, if any, the federal government should play in guiding the utility industry down the path to competition. Some Republican lawmakers, such as House Commerce Committee Chairman Thomas J. Bliley Jr., R-Va., and large utility customers are calling for federal legislation that would require all 50 states to deregulate their utilities.

"While we are in favor of each state enacting a law tailored to its local needs, electricity clearly is an interstate market," said John Anderson, who heads the Electricity Consumers Resource Council, a trade group representing large industrial power users. "Electrons flow across states lines with impunity. We need a federal law that guarantees that all consumers have the right to choose their supplier of electricity." [6]

Deregulation may well lower utility costs for large companies, which will be able to negotiate favorable contracts to buy power in a more competitive environment. But consumer advocates fear that residential customers will face higher utility bills unless federal legislation includes specific language protecting consumers. "We're very focused right now on ensuring that consumers reap the benefits of deregulation as it progresses," says Berman of the Consumer Energy Council. "We definitely are not advocating turning back the clock [to full regulation], but we support legislation and state regulations that would provide consumer protection in a number of ways. A lot of the details of utility deregulation should be left to the states, but there should be some central guidance each state would use to implement its own plan."

The Clinton administration strongly supports utility deregulation and has outlined its own "comprehensive electricity competition plan," which would allow all consumers to choose their power supplier by Jan. 1, 2003. The plan, which Congress is expected to consider this year, would preserve existing consumer protections, promote the use of clean, renewable energy sources to generate electric power and modify federal laws that stand in the way of full competition. The plan grants states considerable flexibility by allowing them to opt in or out of a competitive market.

"At a minimum, Congress needs to act to remove the impediments to retail competition," Richardson said. "In addition, we need to recognize that there are not 50 separate electricity grids out there. Actions that occur in one state can have a profound impact on consumers and utilities located in other states. If state-based competition is to thrive, Congress will need to act."

One of the most controversial issues in the debate over federal legislation involves the time that states would have to open their electricity markets to competition. Some lawmakers, including Senate Energy and Natural Resources Committee Chairman Frank H. Murkowski, R-Alaska, are reluctant to include a "date-certain" deadline for states to complete deregulation. They say such a federal mandate could rush this highly complex process, possibly causing more problems down the road. "I have long said that it is more important to do it right than to do it quickly," Murkowski said at a meeting of the committee last September. "Reliable and reasonably priced electricity is too important to consumers, business, the economy and our international competitiveness to toy with." [7]

However, Berman favors a federally mandated deadline for states to either take steps toward utility deregulation or explain why they will not. "By adopting a federal date-certain provision, Congress could require states to move forward," she says. "They could decide not to deregulate, but they would have to make that decision after a formal process. At the same time, this approach leaves a lot of discretion to the states."

Some experts warn that deregulation may undermine the power grid's reliability. Freed of state requirements and exposed to competition from other providers, utilities may try to save money by drawing down their reserves of surplus electric power, resulting in widespread outages. "I take a pretty dim view of what the regulators and legislators are doing with respect to deregulation," says Chernoff. "Under deregulation, the system is still very reliable, but because the utilities have surpluses and inefficiencies, they have opportunities to lower costs considerably by cutting back on a lot of those surpluses and inefficiencies. Legislators are claiming large savings, but a lot of those savings come from essentially giving away those surpluses. When those surpluses are gone, people are going to find they've got a problem."

Chernoff and others also dispute the claim that deregulation will reduce the cost of electricity. Technological advances are already reducing utilities' operating costs, he says. "Maybe the rate at which they're being accepted has to do with deregulation and restructuring," says Chernoff, who co-authored a recent study on utility deregulation for the American Gas Association. [8] "But it isn't, as claimed, that deregulation is creating these. You'd get the bulk of them without deregulation."

Should the government do more to encourage the use of renewable energy sources?

One of the more innovative elements of President Carter's original energy policy was federal support of research into solar power and other renewable energy sources. Unlike oil and other fossil fuels, renewable sources cannot be readily depleted. Another advantage of renewable, or "green," technologies is that they are far less polluting than fossil fuels. [9]

Federal funding has spurred the development of renewable energy. Once prohibitively expensive, wind farms, biomass generators and solar platforms have become so much more cost-effective that they are within reach of competing with traditional energy sources. But federal support for renewables has waned since the early days, especially during the 1980s, when budget cuts initiated by President Ronald Reagan slashed available funds. While environmental concerns, chiefly the desire to reduce the emission of pollutants and greenhouse gases associated with fossil fuel use, have since renewed support for renewable-energy development, many experts say it should figure even more prominently in any new energy-policy initiatives.

"Most emissions of major pollutants such as sulfur dioxide, nitrogen oxides, particulates and mercury— not to mention carbon dioxide associated with global warming—result from the combustion of fossil fuels," says Carl Weinberg, chairman of the Renewable Energy Policy Project, a Washington think tank that promotes renewable-energy development. "Somehow or other, we've got to get off fossil fuels. We need to shift from coal to natural gas, and eventually to fuels that are produced in cleaner ways."

President Clinton's fiscal 2000 budget proposal includes $2.3 billion for energy efficiency and renewable-energy programs, a $213 million increase over this year's level. Most of the funding would be for research and development of more efficient cars and ways to cut home energy use. But Weinberg says much of the R&D money would be better spent on implementation of existing renewable-energy technologies. "The U.S. government is the largest electricity buyer in this country," he says. "So a good step would be for the General Services Administration to buy 'green' for a portion of its electricity."

But some experts say federal support for renewable energy actually thwarts the overall goal of producing clean and affordable energy. "The whole thing is wrapped up in the notion of creating a level playing field for renewables," Chernoff says. "As with a lot of other uneconomic technologies, the level playing field becomes whatever is necessary to make my technology succeed."

In the case of electricity generation, for example, the Clinton administration's proposal calls for the adoption of "renewables portfolio standards," which would specify the percentage of renewable energy that utilities would be required to purchase from wind farms and other renewable-energy generators.

But favoring renewable energy by setting minimum-use standards is a waste of money, Chernoff says, because it ignores other efficiency improvements, such as new combined-cycle gas plants, which generate additional

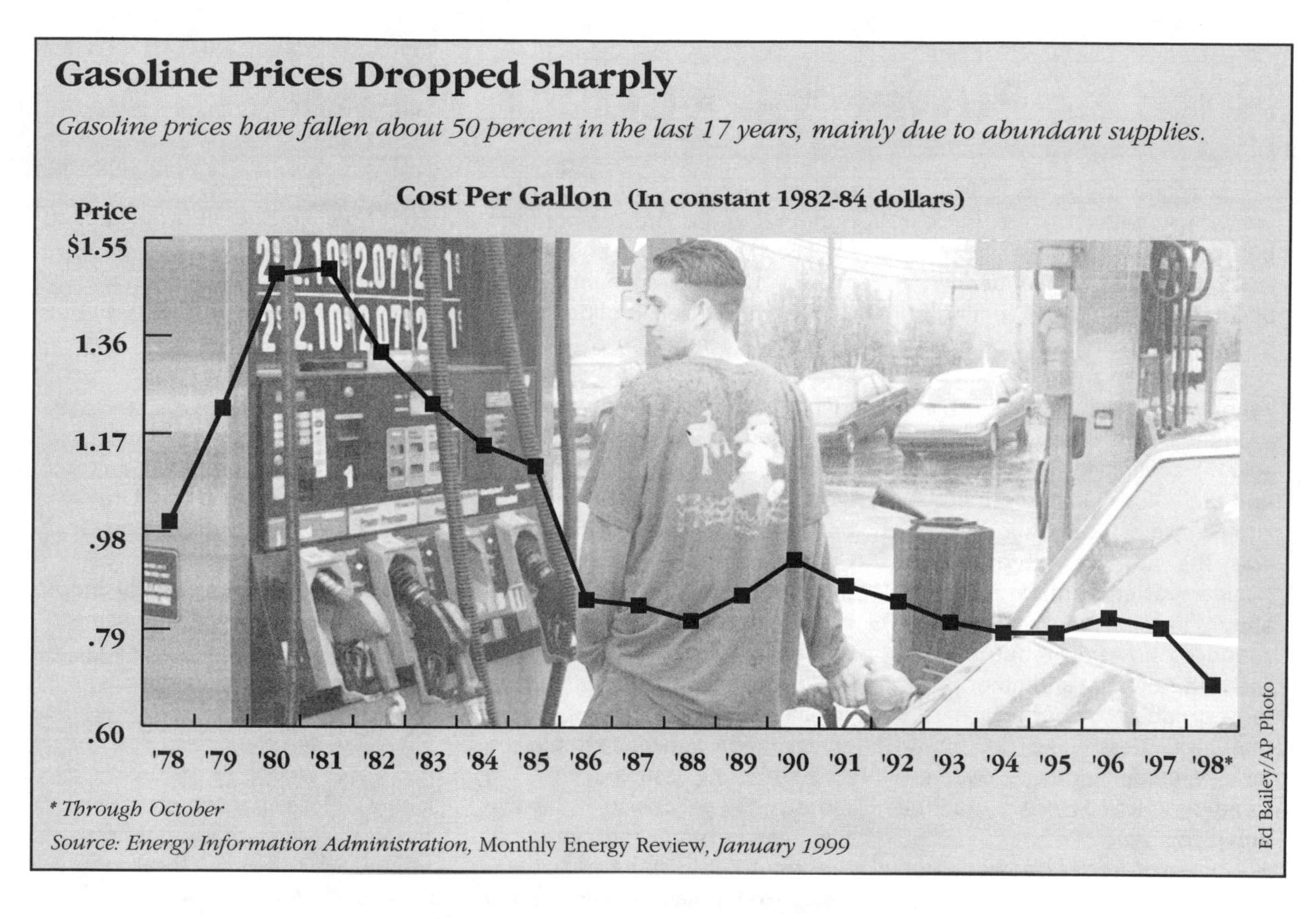

Gasoline Prices Dropped Sharply

Gasoline prices have fallen about 50 percent in the last 17 years, mainly due to abundant supplies.

** Through October*

Source: Energy Information Administration, Monthly Energy Review, *January 1999*

electricity from waste heat and use about half the fuel of traditional boilers fired by gas, oil or coal. "There's nothing in the world that can touch the newest gas-fired power plants, nothing—not coal, not solar, not nuclear," he says. "So anyone who wants to compete against them on the margin with some new technology has to come up with some subsidy or mandate that allows them to compete. The renewables portfolio standards are a way of saying [renewable energy] can't compete in a market context."

Consumer advocates generally support existing policies that fund research and development of renewable energy sources, especially in light of ongoing utility restructuring. "There's so much money involved in utility restructuring that the public interest is getting squeezed as these giant players fight for their economic lives," says Wenonah Hauter, director of Public Citizen's Critical Mass Energy Project. "At the same time, we're not giving enough money to research and development to bring the more sustainable technologies like renewables, or even energy efficiency, to where they need to be."

Background

Oil Crises Fuel Policy

Before the 1970s, lawmakers saw little need for a comprehensive energy policy. For the first half of this century, the United States was the world's biggest oil producer. In the Middle East, the major U.S. oil companies, known as the "Seven Sisters," controlled production, paying royalties to host governments in exchange for the right to extract and market the region's vast oil supplies.

That system fell apart in October 1973, when Arab producing countries imposed an oil embargo on the United States in retaliation for its support of Israel in the Yom Kippur War. The embargo and a subsequent production cutback by OPEC resulted in a sudden increase in oil prices and the first oil shock of the 1970s.

The embargo prompted Congress to approve construction of the $10 billion Trans-Alaska Pipeline, which greatly increased oil supplies in the lower 48 states. The embargo

Nuclear Power vs. Nuclear Waste

The energy crises of the 1970s provided a windfall for the nuclear power industry. In its quest to wean the United States from dependence on foreign oil, the Carter administration's energy policy featured expansion of nuclear energy to meet the country's electricity needs.

Indeed, nuclear energy offered several advantages over both oil and coal—the main fuels then consumed by electric utilities. Not only was the uranium used to fuel nuclear power plants in sufficient supply in the United States, but it produced power without emitting the sulfur dioxide and other pollutants produced by burning coal. Its main weakness—and the industry's eventual undoing —lay in the deadly radioactivity of uranium and its by-products.

Hopes for nuclear power's central role in the country's energy mix ended on March 28, 1979, when a valve burst at the Three Mile Island nuclear plant near Harrisburg, Pa., spilling radioactive water into the building where the reactor was housed. Although the accident caused little apparent damage outside the plant itself, it underscored the dangers inherent in nuclear power and effectively halted further expansion of the industry in the United States. Facilities already licensed or under construction were completed, so the total number of operable reactors in the United States continued to grow in the 1980s. But since peaking at 112 in 1990, the number of reactors in operation has fallen to 103.

With little prospect of further commercial development of nuclear energy in the immediate future, the most pressing issues facing the industry and policy-makers involve the transportation and storage of the mounting volume of nuclear waste. Policy-makers have settled on a single repository for the permanent disposal of all high-level waste from commercial nuclear power plants—a massive facility now under construction at Yucca Mountain, 100 miles northwest of Las Vegas, Nev. President Clinton has proposed spending $234 million—a $71 million increase over 1999—to complete the project.

Yucca Mountain's opening has been repeatedly postponed because of political opposition from Nevada lawmakers, who don't want their state used as the country's nuclear waste dump, and technical concerns about the site's seismic stability and potential for groundwater contamination. The permanent facility is not expected to be ready for use until 2010 at the earliest.

But because nuclear power plants are running out of room for storing their radioactive wastes on-site, utilities and many Republicans support a bill (HR 45) that would open Yucca Mountain immediately for temporary waste disposal, even before the final assessment is complete. Citing environmental risks,

also led to the adoption in 1975 of fuel-efficiency standards that required automakers to increase the gasoline mileage of new cars from an average of 13 miles a gallon to 27.5 miles a gallon by 1985. Also in 1975, Congress established the Strategic Petroleum Reserve to protect the country from future supply disruptions.

The country's second oil shock began during the winter of 1978–79, when the Iranian Revolution disrupted oil flows from the Persian Gulf. The crisis was compounded in 1980 by the outbreak of the Iran-Iraq War. By January 1981, oil prices had reached $34 a barrel, more than 10 times the price in 1972, before the first oil shock began.

In the United States, the supply disruptions and price hikes were felt acutely by a population accustomed to cheap, plentiful gasoline. Long lines at gas stations, coupled with the inflationary impact of higher prices, sparked a public outcry that prompted the Carter administration in the spring of 1977 to launch the country's first sweeping energy-policy initiative. Carter appointed James Schlesinger, a former Defense secretary and Central Intelligence Agency director, to draw up an energy blueprint which the president announced in April 1977 as "the moral equivalent of war."

Carter's energy policy, enacted as the 1978 National Energy Conservation Policy Act, emphasized conservation

President Clinton has threatened to veto the measure.

As an alternative, Energy Secretary Bill Richardson proposed on Feb. 25 that the federal government take legal responsibility for nuclear waste stockpiled at 72 commercial power plants until a permanent repository is ready for use. But Sen. Frank H. Murkowski, chairman of the Energy and Natural Resources Committee, and other lawmakers said that does not alleviate the problem.

Environmentalists and consumer advocates support the president's stand. "Polls show that people have a gut reaction against the transportation of nuclear waste," says Wenonah Hauter, director of Public Citizen's Critical Mass Energy Project. "But somehow that doesn't translate into congressional action on this issue." Passage of HR 45, she says, would result in "100,000 shipments over a period of 30 years moving through about 43 states and exposing 50 million Americans to high-level nuclear waste. It's irresponsible to move this waste until there's a better solution."

Nuclear-waste storage will pose more problems as reactors are decommissioned, or taken out of service. A harbinger of difficulties to come is already apparent in Germany, whose newly elected center-left government has called for the shutdown of all 19 of the country's nuclear power plants. But the plan has run afoul of claims by France and Britain that it violates contracts they have already negotiated for reprocessing spent nuclear fuel from Germany.

In the United States, utility deregulation poses a different set of problems. Public Citizen predicts that introducing competition to the industry will force utilities to close many nuclear plants earlier than they had anticipated. Of the 103 nuclear plants in operation, as many as 90 could be forced to close before their scheduled dates, a process that may cost up to $54 billion. [1]

That prediction seemed to be confirmed last month, when Southern California Edison announced plans to begin the lengthy process of decommissioning one of its three reactors at the San Onofre nuclear complex in Southern California next year, 13 years ahead of schedule. With storage space in short supply for the waste from operating plants, technicians face an even greater task in disposing of an entire reactor. The reactor vessel alone weighs several hundred tons and is so radioactive that only one storage site, located across the country in Barnwell, S.C., is considered secure enough to hold it. But because of its weight, the vessel cannot be transported intact and will have to be cut into pieces before shipping. [2]

[1] Public Citizen, "Stranded Nuclear Waste," Jan. 26, 1999.

[2] See Deborah Schoch, "Idle Reactor to Be Dismantled Early," *Los Angeles Times*, Feb. 19, 1999.

and federal support of research into solar and other renewable energy sources as a means of weaning the country from its dependence on oil, especially imported oil. Property owners received tax credits for insulating their buildings; manufacturers were required to meet energy-efficiency standards for products; consumers paid a tax on "gas-guzzling" vehicles; and oil producers paid a wellhead tax on the crude they extracted in the United States. Carter also ended price controls on domestically produced oil and gradually lifted controls on natural gas prices. [10]

Although OPEC continued to account for more than 60 percent of world oil production outside the Soviet bloc, efforts by the United States and other industrial countries to reduce their dependence on OPEC oil gradually eroded the cartel's ability to control the global market. Oil companies turned their attention toward more reliable sources, such as Alaska and Mexico. Exploration and development of offshore reserves under the British and Norwegian waters of the North Sea were especially fruitful. By June 1975, oil from the huge Brent field and other North Sea deposits had begun flowing, outside the control of OPEC.

Recovery from the first oil shock proved to be short-lived, however. A second shock followed the Iranian Revolution of 1978–79, which swept American ally Shah Mohammed Reza

Pahlavi from power and installed an anti-Western, militant Islamic regime led by the Ayatollah Ruhollah Khomeini. Oil shipments from Iran, then the world's second-largest oil exporter, were disrupted, causing a shortage that boosted the price of a barrel of oil from $13 to $34. It also brought back gas lines in the United States and stirred public condemnation of the oil companies, which were accused of manipulating oil supplies in order to raise prices, and mounting criticism of Carter and his policies.

In April 1979, Carter announced new energy-policy initiatives, notably a lifting of remaining controls on imported oil and a "windfall-profits tax" on "excess" oil-company earnings. He also embraced a controversial proposal to develop "synthetic fuels" from coal and oil shale to reduce the country's dependence on imported oil. Carter's plan, later abandoned as prohibitively expensive, called for producing 2.5 million barrels a day of synthetic fuels by 1990.

Carter's last hopes of salvaging U.S. energy sufficiency—and his presidency—ended on Nov. 4, 1979, when Iranian militants took 50 Americans hostage at the U.S. Embassy in Tehran. The hostages were not released until after Carter's successor, Ronald Reagan, was sworn in on Jan. 20, 1981.

From Crisis to Glut

No sooner did the hostage crisis end than Persian Gulf oil exports slowed to a trickle following the outbreak in September 1980 of the Iran-Iraq War. As each side targeted the other's oil fields, Iran's exports were greatly reduced, while Iraq's virtually ceased. Altogether, the war, which was to last eight years, removed almost 4 million barrels of oil daily from the world market—15 percent of total OPEC output and almost 10 percent of world demand. Oil prices jumped as high as $42 a barrel.

But efforts by the United States and other major oil-importing countries to reduce their consumption were already beginning to take effect. Acting through the Paris-based International Energy Agency, they began coordinating their responses to shortages by drawing down inventories rather than paying premium prices for scarce supplies. Efforts to reduce oil consumption through improved energy efficiency and conservation also helped compensate for the loss of oil exports from Iran and Iraq.

Finally, the quest to diversify sources of foreign oil greatly reduced the dependence of industrial countries on OPEC, as the cartel gradually lost its control over the quantity of oil on the world market. By 1985, OPEC's share of world oil production had fallen to 30 percent. The following year, world demand for oil—and oil prices—had fallen so low that many OPEC producers ignored their production quotas to maintain oil revenues. As oil exports flooded the market, prices collapsed still further. In 1988, U.S. gasoline prices fell from their peak in 1981 to a low of $1.12 a gallon.

The 1980s also saw sweeping deregulation of the U.S. natural gas industry. While price controls had been loosened during the Carter administration, the 1978 Natural Gas Policy Act had left intact a complex system of price constraints based on well location and depth and other considerations. Although Congress initially refrained from revisiting the law, the Reagan administration's strong support for deregulation resulted in a series of rulemaking changes by the Federal Energy Regulatory Commission (FERC) that eroded many of the remaining constraints on prices and marketing practices. In July 1989, Congress removed controls from all remaining domestic production of natural gas. [11]

In 1992, during the administration of President George Bush, the FERC removed more barriers to competition in the industry, resulting in an expansion of U.S. consumption of natural gas and the extensive pipeline system.

Reminder From Iraq

Although the United States and other major consuming nations succeeded in loosening OPEC's grip over oil supplies by shifting to alternative suppliers, the Middle East remained the world's largest source of oil. So when Iraqi forces occupied neighboring Kuwait on Aug. 2, 1990, and the United Nations imposed an embargo on Iraqi oil exports, panic buying ensued, pushing oil prices up again, from $13 a barrel to $40. The Iraqi occupation threatened to disrupt the entire region's exports, and the U.N. in 1991 gave the go-ahead for the United States to lead a coalition of forces in the Persian Gulf War to drive Iraqi troops out of Kuwait.

Although the United States and its allies succeeded in their mission, the gulf war reminded Americans of their continuing vulnerability to Middle Eastern oil supplies. It also prompted Congress to review the energy policy that had been in place with few major changes since the 1970s. The 1992 Energy Policy Act, signed by President Bush on Oct. 24, was a broad measure covering virtually every aspect of energy use. It encouraged conservation, supported renewable energy and alternative fuels and made it easier to build nuclear power plants. The law also opened the door for electric utility deregulation. But lawmakers rejected more radical proposals, notably an energy tax, that might have effectively curtailed demand.

Continued high demand for oil, combined with declining U.S. oil production, has meant that the United States has failed to accomplish a central goal of energy policy since the 1970s—reducing or even eliminating the country's dependence on foreign oil. The U.S. share of total world output of crude oil production fell from 52 percent in 1950 to just 10 percent in 1997. At the same time, oil imports have continued to rise and now account for a record 56 percent of U.S. oil consumption.

Since 1997, the country's persistent vulnerability to future disruptions in world oil supplies has been masked by an oversupply of oil. The financial crisis that swept across Asia in October of that year slowed industrial activity in the region, reducing demand for oil. Unusually warm temperatures in the United States and other consuming countries further dampened demand for oil. For its part, OPEC has been unable to impose the kind of market discipline it used to buoy prices in the 1970s, and oil prices have sunk to their lowest level in decades.

Coal Produces Most U.S. Electricity

More than half the electricity in the United States is generated by burning coal; the second-biggest source is nuclear energy.

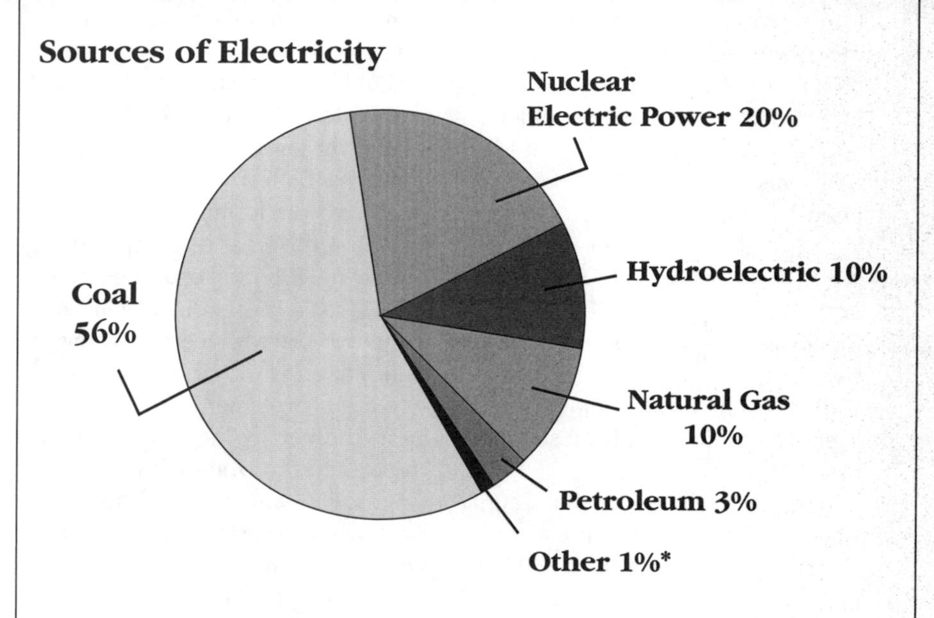

* *Includes geothermal, wood, waste and other sources*

Source: Energy Information Administration, Monthly Energy Review, *January 1999*

Current Situation

Clinton's Energy Policy

Since entering the White House in 1993, President Clinton has presided over the most benign energy environment since the 1970s. Cheap, abundant oil has increased American consumers' buying power and contributed in no small way to the low-inflation, high-employment economy the United States now enjoys. But cheap oil has also made it hard to garner the political support needed to pursue further efforts to reduce the United States' mounting dependence on imported oil.

Clinton sent his most recent energy policy proposal to Congress last spring. Called the "comprehensive national energy strategy," the plan noted important changes in the energy picture since the 1970s, notably deregulation of the oil and natural gas industries, the emergence of international organizations to cope with oil-supply disruptions and heightened awareness of the environmental impact of fossil-fuel use. The plan also called for further cooperation with foreign governments to promote deregulation of energy markets and reduce greenhouse emissions as well as increased funding of programs to develop clean and efficient energy technologies.

Clinton's fiscal 2000 budget calls for increased spending for renewable-energy technology and energy-efficiency programs. Some of the administration's specific proposals include development of an 80-mile-per-gallon car, reducing home-energy use by half by 2010 and cuts in energy use by government agencies and industry. Clinton also calls for increased funding of natural-gas research and nuclear energy programs as a way to reduce greenhouse gas emissions by electric-power plants.

Do Diplomacy and Energy Security Mix?

Ever since the Arab oil embargo touched off the first energy crisis in the 1970s, U.S. foreign policy and energy needs have been closely intertwined, and often in conflict.

Ensuring access to the Middle East's vast reserves of oil has long been a focus of U.S. foreign policy in the region. Even after the Arab-led Organization of Petroleum Exporting Countries (OPEC) orchestrated an oil embargo and successive price hikes in the 1970s, the United States has maintained friendly relations with most of its member governments.

It is no coincidence that Saudi Arabia, the country with the world's largest oil reserves and OPEC's most influential member, is a key U.S. ally in the region. When Iraq invaded Kuwait on Aug. 2, 1990, the U.S.-led coalition of United Nations forces ousted the invaders with the formal objective of protecting Kuwait's sovereignty, but also with the tacit mission of preventing Saddam Hussein's hostile government from disrupting oil exports from the Persian Gulf.

But U.S. interests in the Caspian Sea region, the site of potentially vast reserves of crude, are more complex. Concerns over tensions in the region, formerly part of the Soviet Union, and competition among neighboring countries hoping to cash in on the coming oil bonanza, often place U.S. foreign policy at odds with its energy strategy. A major element of that strategy is to diversify the sources of oil imports to the United States away from OPEC and the Middle East. And while access to Caspian oil would help meet that goal, it is at odds with other foreign policy interests.

A key source of the Caspian's oil riches is Azerbaijan. But in 1992, acting on reports of human rights violations by Azerbaijan's government against Armenian rebels in the breakaway province of Nagorno-Karabakh, Congress imposed sanctions against Azerbaijan. The sanctions, strongly advocated by the influential Armenian Assembly of America, suspended U.S. assistance programs to the country just as its value as a potential oil supplier was becoming apparent.

Foreign policy interests also complicate the technical difficulties involved in transporting Caspian crude to market. The region is landlocked, and the oil must travel thousands of miles to reach seaports before it can be loaded onto tankers and shipped. Because rail transport is expensive and environmentally risky, oil pipelines offer the only viable alternative. But every potential pipeline route poses political drawbacks. The most direct route to open water would pass through Iran to the Persian Gulf. The United States opposes that option, however, because it would defy existing U.S. sanctions against Iran for its alleged role in

Some critics oppose such traditional efforts to encourage alternative energy sources. "The programs that the government is subsidizing or mandating with respect to alternative fuels and new types of vehicles are really just backwards," says Chernoff at Science Applications International Corp. "If you want to get more efficient usage of fuel, you shouldn't be switching fuels, you should be raising the price of the one you want to use less of. In other words, you should apply a gasoline tax." There is no significant political support, however, for raising the federal gas tax, now 4.3 cents a gallon.

Other critics of Clinton's energy policy fault the administration for failing to enforce existing energy requirements, including a 1994 executive order by Clinton himself that government agencies cut energy use by 20 percent over 1985 levels in 2000 and by 30 percent in 2005. According to the Alliance to Save Energy, taxpayers could save $1 billion each year and see a significant reduction in air pollution and greenhouse gas emissions if the federal government tried to meet the requirements, such as by switching to energy-efficient halogen lights.

"Reducing the federal government's massive energy waste offers enormous opportunities to save taxpayers billions of dollars for decades to come and improve the environment," said Sen. Jeff Bingaman, D-N.M., senior Democrat on the Senate Energy and Natural Resources Committee and the alliance's co-chairman. "We're starting to make progress, but there's no excuse for this much waste when leading companies in the U.S. energy-efficiency industry are willing to provide the money for improvements at no up-front cost to taxpayers." [12]

supporting international terrorism. An Iranian pipeline would also be vulnerable to sabotage in that troubled region. [1]

Although the U.S. sanctions apply to foreign as well as U.S. companies, Secretary of State Madeleine K. Albright waived them last May for three European oil companies that had invested in Iran in defiance of U.S. policy. American oil companies have since complained that the sanctions serve merely to erode American competitiveness in the region. Those complaints seem certain to grow louder after the newly formed British-American oil giant, BP Amoco PLC, recently announced plans to develop three major fields in southern Iran. [2]

Other relatively direct routes would pipe Caspian oil to the Black Sea. But these would place pipelines dangerously close to Nagorno-Karabakh and other troubled areas of Azerbaijan, Armenia, Georgia and southern Russia. The Clinton administration is pushing oil companies and governments to choose what it considers to be a safer pipeline route across Turkey, a NATO ally, to the Mediterranean Sea. But oil companies, which must foot the bill for pipeline construction, are reluctant to opt for that route, which would be about twice as long as the other options, especially at a time when low oil prices have drained corporate revenues.

The current oil glut and collapse in oil prices pose another problem for U.S. policy-makers. While American consumers are enjoying cheaper gasoline, countries that depend on oil exports for their economic well-being face declining living standards, which is threatening political stability in some cases. This is especially true in economically devastated Russia, which is trying to lure foreign investors into its crippled oil industry.

Another sign of cheap oil's adverse impact on producers came last fall, when the Saudi government invited seven American oil companies to discuss the possibility of investing in the kingdom's oil and gas fields for the first time since driving them out in the early 1970s.

"People aren't paying attention to the systematic impact of very low oil prices on the oil-producing countries around the world," says Daniel Yergin, chairman of Cambridge Energy Research Associates Inc. and author of *The Prize*, an award-winning history of the oil industry. "Yet this is one of the major reasons for the economic collapse in Russia. We need to be thinking about the foreign policy and stability impacts [of low prices] on other countries."

[1] For background, see Mary H. Cooper, "Oil Production in the 21st Century." *The CQ Researcher*, Aug. 7, 1998, pp. 673–696, and David Masci, "Reform in Iran," *The CQ Researcher*, Dec. 18, 1998, pp. 1110–1133.

[2] See David B. Ottaway and Martha M. Hamilton, "BP Amoco Seeks to Drill in Iran," *The Washington Post*, Jan. 30, 1999.

Utility Deregulation

A key element of Clinton's energy policy is promotion of electric-utility restructuring, outlined in March 1998 in his comprehensive national energy strategy. Dismantling utility regulations, the Energy Department estimates, would save consumers $20 billion a year on their electricity bills—the equivalent of $232 a year for a typical family of four. The plan would make it federal policy to allow all consumers to choose their power suppliers by Jan. 1, 2003, while preserving the regulated system's reliability and promoting the use of renewable-energy sources for power generation.

In supporting deregulation, says Yergin of Cambridge Energy Associates, Clinton is merely recognizing a global shift away from government intervention in energy markets. "The world is changing its mind about what government should do, and the energy industry is case study No. 1," he says. "We're really moving away from the legacy of the New Deal to less confidence in the ability of government to keep up with and manage markets and greater confidence in the ability of markets to work. That's the Zeitgeist of the age."

The 1992 Energy Policy Act and subsequent rulings by the Federal Energy Regulatory Commission started the process of electric-utility deregulation by allowing utilities greater leeway in choosing the generators of the power they sell to their customers. In 1996, California adopted the first state law opening the retail side of the industry to competition as well.

Even before Clinton lent his support to the process, utility deregulation was well under way. Protected as a "natural monopoly" under a regulatory framework established

Global Warming Boosts Renewable Energy

For more than 100 years, scientists have suspected that burning fossil fuels—especially coal and oil—causes a gradual and potentially devastating warming of Earth's atmosphere. According to the global warming theory, carbon dioxide and other gases released during fuel combustion trap solar heat in the atmosphere in much the same way as the glass roof and walls of a greenhouse. Even a slight rise in global temperature would melt polar ice, raising sea levels in some of the world's most populated areas and causing widespread flooding and disruptions in weather patterns and agricultural production as well as the proliferation of infectious diseases. [1]

Mounting evidence in support of that theory led President Clinton and 175 other world leaders meeting in Kyoto, Japan, in December 1997 to sign an agreement calling for reduced emissions of greenhouse gases. The Kyoto Protocol would require the United States—the world's biggest consumer of fossil fuels —to reduce its greenhouse gas emissions to 7 percent below 1990 levels by 2012. Other industrial countries would also have to meet emissions targets, but developing countries were exempted from legally binding mandates on the ground that they have contributed relatively little to the problem.

Criticism of the protocol runs so high in the Senate, which must ratify it by a two-thirds majority, that the White House has not even submitted it for consideration. Some senators say the scientific evidence in favor of global warming is not strong enough to justify the sweeping and costly reductions in coal and oil consumption required under the protocol. Critics also charge that the protocol unfairly penalizes industrial countries by exempting rapidly developing countries such as China and India from legally binding mandates to curb fossil fuel use. The Senate adopted a resolution sponsored by Sen. Robert C. Byrd, D-W.Va., calling for a reversal of the exemption for developing countries.

While disputes over the timing and extent of emissions reductions make Senate ratification of the agreement unlikely anytime soon, many environmentalists and energy experts have welcomed the debate as an essential first step toward dealing with global warming. "Unfortunately, Kyoto has become a lightning rod for this issue," says David M. Nemtzow, president of the Alliance to Save Energy, a Washington, D.C., coalition that promotes energy efficiency. "Congress is behind in the climate-change debate. In the real world, the debate has shifted from the science of climate change to the costs of responding to it."

Even without Senate ratification of the Kyoto Protocol, concerns about the environmental impact of energy use are boosting support for the development of renewables and other less-polluting energy sources and for better efficiency of energy use. According to the Worldwatch Institute, global wind-energy generating capacity reached 9,600 megawatts in 1998, double the capacity in place three years earlier. [2] And the Washington-based Renewable Energy Policy Project predicts that an additional 10,000 megawatts of wind energy could be produced over the next decade. [3]

Meanwhile, even some businesses that were among the most vocal opponents of the Kyoto Protocol in 1997 are taking steps on

Digital Stock

Global wind-energy generating capacity has doubled in the last three years.

their own to curb greenhouse gas emissions. Two major oil companies—British Petroleum PLC and Royal Dutch/Shell Group of Cos.—announced voluntary emissions reductions last fall. United Technologies Corp., which builds aircraft engines and air conditioners, said it would reduce its energy use by 25 percent over the next decade. [4]

These steps are welcome news to environmentalists and many energy analysts alike. "Countries can sign the Kyoto Protocol and not do anything, or they can do something and not sign up," Nemtzow says. "As a practical matter, we want people to do the right thing, whether they sign the protocol or not."

[1] For background, see Mary H. Cooper, "Global Warming Update," *The CQ Researcher,* Nov. 1, 1996, pp. 961–984.

[2] Christopher Flavin, "Wind Power Set New Record in 1998: Fastest-Growing Energy Source," Worldwatch Institute, Dec. 30, 1998.

[3] Jamie Chapman and Steven Wiese, "Expanding Wind Power: Can Americans Afford It?" Renewable Energy Policy Project, October 1998.

[4] See Martha M. Hamilton, "Firms Warm Up to Climate Treaty," *The Washington Post,* Nov. 2, 1998.

by the 1935 Public Utilities Holding Company Act, the industry began to face limited competition as a result of the 1970s energy crises. The 1978 Public Utility Regulatory Policies Act required utilities to purchase some of their power from outside sources. This initial change in traditional electricity regulation was amplified by the 1992 Energy Policy Act, which allowed new, unregulated entities to generate and sell electricity to utilities, and by subsequent rulings by FERC. [13] These steps allowed utilities greater freedom to choose the providers of power that they distribute to customers.

Assessing the Risks

As the federal regulatory requirements were loosened, states began to deregulate their utilities. In 1996, California adopted the first state law opening the retail side of the industry to competition. Since then, about 20 states have followed California's lead by dismantling regulations or setting up power commissions to study ways to change industry regulations.

The incentive to deregulate utilities is enormous. Of the approximately $500 billion spent each year for energy in the United States, $200 billion is for electrical power. "That's more than this nation spends for all long-distance, local telephone, cell paging, Internet and other telecommunications services put together," says Nemtzow at the Alliance to Save Energy. "For this reason, there are enormous stakes in changing the federal ground rules."

Deregulation has not proceeded smoothly or uniformly, however. Last November, California voters rejected a controversial ballot initiative that would have overturned a provision of the deregulation law making consumers shoulder the costs of utility plants that are no longer economically viable under deregulation. But a consumer backlash against deregulation may be brewing in California, setting an example for other states as they contemplate utility restructuring. When offered the chance to switch electricity providers in January, only a small percentage of California customers opted to do so.

"Just as in the early days of phone deregulation, people are just too confused to make informed choices," says consumer advocate Berman. "With the fast pace of mergers, many of the names of utilities are changing. How do you know who is a fly-by-night operator and who is reliable?"

Some experts worry that by forcing utilities to compete for customers, deregulation will compromise the U.S. electricity grid's reliability. In view of last summer's brush with a major outage in the Midwest, Chernoff worries that deregulation will result in far more serious power interruptions.

"Under regulation, all the utilities in a region have a requirement for certain levels of surplus capacity, and the utilities operate cooperatively and collaboratively to make sure that if one is deficient someone else can cover for it," he says. But over the last five or 10 years the utilities have been cutting back on their reserve margins in anticipation of deregulation. "That frees up a lot of capacity, which enables them to lower costs and prices," he says. "It also makes the legislators who are beating the drums for deregulation look good. But at some point we're going to run into the fact that the system is operating on a much tighter margin."

Proponents of utility deregulation predict that the risk of power outages will disappear once the transition from regulated to competitive markets for electricity is complete. "As the industry continues to evolve from the old model of regulated monopolies to a new model based on competition, customer choice and free markets, we are obviously going to experience a period of significant change and adjustment," James L. Turner, vice president of Cinergy Corp., a Cincinnati, Ohio, utility, told the Senate Energy and Natural Resources Committee last September. "The price spikes in the Midwest this past summer are a manifestation of such adjustment, but more than anything they suggest the need to finish the job of opening not just wholesale, but retail, markets to full and fair competition." [14]

Another concern is that while many American consumers and businesses will pay less for power in a deregulated market, customers in states where electricity prices have been kept low under regulation will face steep price hikes. States that produce power with hydroelectric dams or coal-fired plants are expected to face the biggest cost increases when they begin to replace these older facilities with new, state-of-the-art plants. "A lot of the new capacity, even low-cost capacity," Chernoff says, "is going to cost more on average than the old capacity that was bought 30 or 40 years ago when it cost a tiny fraction of what it costs today."

Apart from price differences among states, consumer advocates fear that residential customers will suffer in a competitive marketplace for electricity because they lack the economic clout that will enable large industrial users to negotiate favorable rate structures. "None of the bills passed at the state level provide enough protections for consumers in the long run," says Hauter of Public Citizen. The group supports federal legislation to protect residential customers and small businesses from rate hikes and ensure that renewable energy plays a significant role in power generation. In the absence of such federal guidelines, Hauter says, "Deregulation is moving the industry in the direction of larger, less accountable organizations."

As lawmakers consider proposals for federal guidelines to avoid inequities and conserve the grid's reliability, they are also coming under pressure from power generators that are poised to enter the deregulated market. "A retreat to 're-regulation' would be a mistake," Susan Tomasky, general counsel of American Electric Power, an Ohio utility, told the Energy Committee. "More reliance on market forces, not less, is needed to assure that supply and demand for electricity remain in balance." [15]

Oil Industry Mergers

Just as utility restructuring is radically changing the way Americans buy electricity, low oil prices are leaving their mark on the other major component of the U.S. energy market, the oil industry. The price slump that has helped enrich consumers and industrial users has wreaked havoc on oil-company earnings, prompting industry calls for help from the federal government.

The Clinton administration responded to one of their demands on Feb. 11, when it announced plans to move 28 million barrels of domestic oil into the Strategic Petroleum Reserve in an effort to halt the erosion of oil prices. Beginning in April, the government plans to transfer the oil, worth about $330 million at current prices, at the rate of 100,000 barrels a day. To avoid spending tax revenues on the transfer, the government will take the oil in lieu of royalties producers must pay on oil extracted on federal lands. "We want to help our domestic industry," Energy Secretary Richardson said. "They are hurting. We recognize that." [16]

While domestic producers are clamoring for federal assistance to help them survive the price slump, big oil companies are trying to cut costs by buying out the competition, both in the United States and overseas. In December, French oil giant Total S.A. announced plans to acquire Belgium's Petrofina S.A., in what analysts described as the first step in a consolidation of the European oil industry. [17] On Dec. 31, the Federal Trade Commission (FTC) approved the buyout of Amoco Corp. by British Petroleum PLC after BP agreed to sell a number of gas stations in the United States. The new company, BP Amoco PLC, is now one of the world's largest oil companies. Mobil, which reported a $12 billion drop in revenue last year, agreed last Dec. 1 to be purchased by Exxon. If approved, the $75 billion deal would constitute the biggest merger in U.S. history and create the world's largest oil company.

Consumer advocates worry that big oil mergers will result in higher energy prices for consumers down the road. "We know that a consolidation of this nature leads to more consolidation," says Hauter of Public Citizen. "Oil prices are at record lows, but anyone who has followed the history of the oil industry knows there's been collaboration—some might say conniving—among these companies to try to keep production down and prices high." Hauter and other consumer advocates are calling for congressional hearings into the long-term impact of oil-company mergers and caution on the part of antitrust regulators in approving such megadeals.

The FTC has yet to complete its antitrust review of the Exxon-Mobil merger and may require the companies to sell refineries and sever contracts with more than 1,000 gas stations in the Northeast, where they have a large share of the market, before it approves the merger. [18]

Oil company representatives say the mergers promise to yield benefits to consumers over the long run. "The mergers that have occurred and are currently being considered fall well below the level of market concentration the government has traditionally been worried about," says the API's Caveney. "The very reason that mergers occur is to create efficiencies. So we see them as very positive for the consumer because the new companies will be able to reduce their operating costs even further."

Outlook

Overhaul Unlikely?

For all the changes taking place in the U.S. energy industry, there is little political momentum to undertake a major revision of the energy-policy framework that has been in place since the 1970s. "There are a lot of energy issues that demand attention, but the congressional agenda is modest," says Nemtzow of the Alliance to Save Energy. "I don't think that the 106th Congress will pay a great deal of attention to them."

Mounting losses in "oil-patch" states have placed aid to the domestic oil industry near the top of the energy-policy agenda. In January alone, 11,500 more oil-industry jobs disappeared, the most for any month since the mid-1980s, bringing to 42,000 the total number of jobs lost to low oil prices since the slump began.

"Our oil industry is in serious trouble," said Sen. Murkowski. "And if we don't do something, it will only get worse." He proposes a number of measures to boost domestic production, such as increased funding for petroleum research and development and more lenient rules for calculating royalties that oil companies pay the government to pump oil on federal land.

Far more controversial are proposals supported by Murkowski, Hutchison and some other lawmakers to give

FOR MORE INFORMATION

Alliance to Save Energy, 1200 18th St. N.W., Suite 900, Washington, D.C. 20036; (202) 857-0666; www.ase.org. A coalition of government, business, consumer and labor leaders concerned with increasing the efficiency of energy use.

American Petroleum Institute, 1220 L St. N.W., Washington, D.C. 20005; (202) 682-8100; www.api.org. The leading membership organization for the oil and natural gas industries provides information on consumption, prices and oil-related policy issues.

Energy Information Administration, U.S. Department of Energy, Washington, D.C. 20585; (202) 586-8800; www.eia.doe.gov. The Energy Department's information branch provides data on consumption, imports and prices of fuels consumed in the United States.

Critical Mass Energy Project, Public Citizen, 215 Pennsylvania Ave. S.E., Washington, D.C. 20003-1155; (202) 546-4996; www.citizen.org/CMEP. This public-interest group promotes energy efficiency and renewable-energy technologies and opposes further development of nuclear energy.

Renewable Energy Policy Project, 1612 K St. N.W., Suite 410, Washington, D.C. 20006; (202) 293-2898. This nonpartisan think tank promotes the development of non-polluting, renewable-energy sources as a growing part of the U.S. energy mix.

Edison Electric Institute, 701 Pennsylvania Ave. N.W., Washington, D.C. 20004; (202) 508-5000; http://www.eei.org. The leading membership association representing investor-owned electric power companies and electric utility holding companies provides information on deregulation, renewable energy and other issues involving the utility industry.

the oil industry greater access to federal lands. Oil companies have long lobbied to gain access to areas that are currently off-limits, notably the Arctic National Wildlife Refuge. According to the U.S. Public Interest Research Group (USPIRG), ARCO, BP-Amoco, Exxon and Chevron have made more than $8.7 million in campaign contributions to further that goal.

"The coastal plain should be spared the fate of Prudhoe Bay and should be designated as wilderness," said USPIRG's Athan Manuel. "An area as unique and pristine as the Arctic Refuge should be preserved, not plundered." [19]

While proposals to help the oil industry cope with low oil prices are likely to gain support in the 106th Congress, the main focus of legislative proposals to change energy policy is over what role, if any, the federal government should play in electric-utility deregulation. Analysts differ over the likelihood that Congress will stipulate a deadline for the completion of state deregulation. "I think the date-certain approach is the right way to go," attorney Zausner says. "It worked brilliantly with natural gas, and I believe it would work brilliantly for utilities, too."

But lawmakers are divided on this issue. House Commerce Committee Chairman Bliley "is a true believer in date-certain," Zausner says. "But my perception is that the politics for it are not there. That's certainly true in the Senate, where it will be approved over Murkowski's dead body."

The political logjam could well break, Zausner predicts, if a serious power outage occurs this summer. "I think there are going to be reliability incidents because of the way deregulation is happening," he says. "We got damned close to that last year, and the longer we have this berserk, hodgepodge system, the more likely we are going to see it again."

Barring a serious blackout, however, many analysts agree that the very complexity of the energy debate and the myriad interests that have stakes in its outcome stymie progress in the legislative arena. "This is a high-stakes game in an issue that is not currently facing a crisis," Nemtzow says. "Like telecommunications deregulation and clean-air legislation, utility deregulation will take a long time—and multiple Congresses—to tackle." ❖

Thought Questions

1. If you were the President's energy secretary what recommendations would you make for reducing our nation's dependence on imported oil? What political problems do you foresee for your recommendations?
2. Should Congress legislate higher auto fuel efficiency standards? Why or why not?
3. Should Congress legislate a higher gas tax to encourage Americans to conserve energy? Why or why not?
4. The state of Nevada has never had a nuclear power plant, yet it is the site of the proposed Yucca Mountain nuclear waste repository. Many Nevadans consider this unfair, whereas many other Americans think that Nevada would be the perfect site to house nuclear waste because of its low population density. Should Nevada be compelled to house nuclear waste from other states, or should it have the right to decline? What other possibilities are there for dealing with nuclear waste?
5. Critics of the Kyoto Protocol to curb greenhouse gas emissions think that it is unfair that developing nations would be exempt from greenhouse gas reductions while the United States and other industrialized nations would be required to reduce their emissions to pre-1990 levels. Do you agree with this criticism? Why or why not?

Notes

[1] For background, see Mary H. Cooper, "Oil Production in the 21st Century," *The CQ Researcher*, Aug. 7, 1998, pp. 673–696.
[2] Robert Gehrke, "Conference Members Lament Oil Industry Woes," The Associated Press, Dec. 7, 1998.
[3] Dewey F. Bartlett Jr., "Shame on Those Who Ignore U.S. Oil Industry's Plight," *The Daily Oklahoman*, Dec. 4, 1998.
[4] U.S. Department of Energy, *The Changing Structure of the Electric Power Industry: Selected Issues*, 1998. For background, see Kenneth Jost, "Restructuring the Electric Industry," *The CQ Researcher*, Jan. 17, 1997, pp. 25–48.
[5] Richardson addressed a meeting of the Edison Electric Institute in Scottsdale, Ariz., on Jan. 8, 1998.
[6] Quoted in Electricity Consumers Resource Council, "Energy: The Power to Choose," an advertising supplement published in *The New York Times*, Jan. 18, 1998.
[7] Murkowski spoke at a committee hearing on Sept. 24, 1998.
[8] Harry Chernoff and Gabriel Sanchez, Science Applications International Corp., "The Impact of Industry Restructuring on Electricity Prices," July 1998.
[9] For background, see Mary H. Cooper, "Renewable Energy," *The CQ Researcher*, Nov. 7, 1997, pp. 961–984.
[10] Information in this section is based in part on Daniel Yergin, *The Prize* (1991).
[11] See Pietro S. Nivola, "Gridlocked or Gaining Ground: U.S. Regulatory Reform in the Energy Sector," *The Brookings Review*, summer 1993, pp. 34–41.
[12] Bingaman spoke at a news conference on Nov. 10, 1998. See "So Much for Climate Change; Federal Government Energy Waste Costs Taxpayers $1 Billion Annually," *PR Newswire*, Nov. 10, 1998.
[13] For background, see Amy Abel, "Electricity Restructuring Background: The Public Utility Regulatory Policies Act of 1978 and the Energy Policy Act of 1992," *CRS Report for Congress*, Congressional Research Service, May 4, 1998.
[14] Turner testified before the Senate Energy and Natural Resources Committee on Sept. 24, 1998.
[15] Tomasky testified before the Senate Energy and Natural Resources Committee on Sept. 24, 1998.
[16] Quoted by Warren Brown, "U.S. to Transfer Oil to Bolster Prices," *The Washington Post*, Feb. 12, 1999.
[17] See John Tagliabue, "Total Says It Will Pay $13 Billion for Petrofina," *The New York Times*, Dec. 2, 1998.
[18] See John R. Wilke and Steve Liesman, "Exxon, Mobil May Be Forced Into Divestitures," *The Wall Street Journal*, Jan. 20, 1999.
[19] Athan Manuel, "No Refuge: The Oil Industry's Million Dollar Campaign to Open Up the Arctic," *U.S. Public Interest Research Group*, February 1999, p. 11.

Chronology

1970s *The Arab oil embargo prompts the United States to adopt its first sweeping energy policy.*

October 1973
Oil prices soar from around $5 a barrel to above $17 after the Organization of Petroleum Exporting Countries (OPEC) imposes an embargo against the United States for its support of Israel.

1975
The Strategic Petroleum Reserve is created to protect the United States from disruptions in oil supplies. Congress sets fuel-efficiency standards for cars.

April 1977
President Jimmy Carter announces a sweeping energy strategy that he calls "the moral equivalent of war."

1978
The National Energy Conservation Policy Act emphasizes conservation and federal support of research into alternative-energy sources to help reduce U.S. dependence on foreign oil. The Public Utility Regulatory Policies Act begins the process of deregulating the electric-utility industry by requiring utilities to buy some of their power from outside sources.

December 1978
The Iranian Revolution disrupts oil supplies from the Persian Gulf, causing a second oil shock and increasing inflation in industrial countries.

March 28, 1979
An accident at the Three Mile Island nuclear power plant in Pennsylvania dashes plans to have nuclear energy play a prominent role in the country's energy mix.

April 1979
President Carter expands energy policy to include a "windfall-profits tax" on "excess" oil-company earnings and an ill-fated plan to develop "synthetic fuels" from coal and shale oil.

Nov. 4, 1979
Iranian militants take 50 Americans hostage at the U.S. Embassy in Tehran.

1980s *Oil prices ease as consumers shift to non-OPEC suppliers.*

1980
The Iran-Iraq War breaks out, further disrupting Persian Gulf exports and raising oil prices to around $34 a barrel.

July 1989
Congress removes controls from domestic production of natural gas, furthering the deregulation that began in the 1970s.

1990s *U.S. dependence on foreign oil continues to grow.*

Aug. 2, 1990
Iraqi forces invade Kuwait, prompting a rise in oil prices. The Persian Gulf War ends the occupation the following year, when a U.S.-led United Nations force ousts the invaders.

Oct. 24, 1992
President George Bush signs the Energy Policy Act, which addresses a broad array of energy issues, including further steps to deregulate electric utilities, but stops short of imposing an energy tax to curb consumption.

1996
California adopts the first state law allowing consumers to choose their suppliers of electrical power.

December 1997
President Clinton signs the Kyoto Protocol calling for reductions in emissions of greenhouse gases to slow global warming.

March 25, 1998
The Clinton administration presents its "comprehensive electricity competition plan," calling for cooperation with other countries to speed electric utility deregulation and increased funding of programs to develop clean and efficient energy technologies.

Dec. 1, 1998
Exxon Corp. announces plans to acquire Mobil Corp. If approved, the merger would be the biggest in U.S. history.

Dec. 31, 1998
British Petroleum PLC and Amoco Corp. win approval to merge. The new company, BP Amoco PLC, is one of the world's biggest oil conglomerates.

Feb. 11, 1999
The Clinton administration announces plans to move 28 million barrels of domestic oil into the Strategic Petroleum Reserve in an effort to halt the erosion of oil prices.

At Issue:

Will oil-company mergers lead to higher fuel prices?

WENONAH HAUTER—Director, Public Citizen's Critical Mass Energy Project

From a statement issued Dec. 8, 1998.

The merger between Exxon and Mobil creates the largest corporation in the world and will lead to non-competitive practices that will harm consumers for decades to come. The trustbusters were right 90 years ago during the Teddy Roosevelt era when they broke up Rockefeller's Standard Oil monopoly into more than 30 companies. Lack of competition was bad for consumers then, and it will be bad for consumers today.

Just because oil prices are at their lowest point since the Great Depression in the 1930s doesn't mean they will not rise in the future. Today's prices are the result of a world oil glut exacerbated by the economic woes in Asia, the unusually warm weather in the United States, reduced consumption of heating oil in Europe because of a transition from oil to natural gas for heating, more oil production in places like West Africa and the partial re-entry of Iraq into world markets.

Currently, there is competition in refining and marketing, and that is also playing a role in lower prices. . . . When Shell and Texaco combined their U.S. marketing operations, that had an impact on prices in places like California where some competition was eliminated. Prices are higher in Southern California than in the Northeastern United States because there is less competition. Prices are lower in Los Angeles than in San Diego because there is more retail competition.

If the first and second-largest oil companies in the United States are allowed to merge, there will be problems for consumers down the road. These companies were once fierce competitors, and this will eliminate significant competition. Also, a merger of this magnitude will lead to other mergers and acquisitions in the industry. We could be left with just a handful of big oil companies, portending less competition at every stage of production.

Much of the oil industry is vertically integrated. That means one company owns and operates facilities in all phases of the production and delivery of a product. As the oil industry continues to consolidate, there will be less competition in extracting and refining oil. . . .

This merger is good for the CEOs and Wall Street, because executives and some investors will make a fortune in the short term. It is not good for consumers in the long term, for the thousands of workers who will lose their jobs or for the environment. The oil industry is a profitable business today. It is abhorrent that a few individuals should reap a huge windfall at the expense of consumers.

Yes

AMERICAN PETROLEUM INSTITUTE

From an API briefing, Dec. 3, 1998.

Americans look to oil and natural gas for more than three-fifths of the energy they rely upon at work, at home and on the road. They will continue to receive the benefits that come from having a variety of oil companies vigorously competing for their business. Petroleum is still one of the least concentrated major U.S. industries, even after taking into account the most recent merger announced by Exxon and Mobil. That merged [corporation] would account for less than 12 percent of U.S. refining capacity and about 13 percent of U.S. gasoline marketing (about the same as BP Amoco and from the merger of Texaco's and Shell's U.S. downstream operations).

By comparison, here are the market shares accounted for by the largest U.S. company in various other industries: 38 percent in automobile manufacturing; 28 percent in freight rail; 38 percent in aluminum; 48 percent in tobacco products; 45 percent in overnight delivery services; and 44 percent in soft drinks.

In addition, the market for oil is global rather than confined within the borders of any one nation, even a nation as large as the United States. On the global market, even a company as large as the new Exxon Mobil Corporation would account for 7 percent of refining capacity, 11 percent of marketing and 1 percent of crude oil reserves. World crude oil reserves now represent nearly a half-century of supply at current rates of consumption. World oil prices, adjusted for inflation, are near historic lows largely because supplies have grown so large relative to demand.

Historically low prices for the products they sell are putting unprecedented pressure on oil companies to become as efficient as possible in order to remain profitable and capable of generating the capital needed to compete and supply energy products in the future. The continual pressure to become more efficient and cut costs has contributed to the low prices for refined petroleum products that consumers now enjoy.

For instance, motor gasoline is the oil industry's single most important product. Today, oil companies must try to make a profit on gasoline after receiving only about 27 percent of the revenue per gallon (after adjusting for inflation) that they received in the early 1980s. . . .

In November 1998, a gallon of unleaded regular sold for an average of 98.9 cents a gallon. After deducting 43 cents for taxes, 55.9 cents remained to cover all the costs of making gasoline available to motorists for purchase at the pump. This 55.9 cents is about 27 percent of the 206.9 cents (after taxes) paid by motorists in 1981.

No

Bibliography

Selected Sources Used

Books

Yergin, Daniel, and Joseph Stanislaw, *The Commanding Heights: The Battle Between Government and the Marketplace That Is Remaking the Modern World*, Simon & Schuster, 1998.
The dismantling of governments' role in the marketplace is changing the way economies work around the world. In the United States, policymakers are grappling with this process in the ongoing deregulation of the electric utility industry.

Yergin, Daniel, *The Prize: The Epic Quest for Oil, Money & Power*, Simon & Schuster, 1991.
This Pulitzer Prize-winning study traces the history of the global oil industry and its impact on energy consumption in the United States.

Articles

Bremmer, Ian, "Oil Politics: America and the Riches of the Caspian Basin," *World Policy Journal*, spring 1998, pp. 27–35.
Securing access to the Caspian Sea region's oil reserves has supplanted containment of the Soviet bloc as the main focus of U.S. foreign policy. But the author questions policy-makers' ability to attain that goal without further destabilizing this volatile region.

Cooper, Richard N., "Toward a Real Global Warming Treaty," *Foreign Affairs*, March/April 1998, pp. 66–79.
Developing countries are reluctant to accept limits on emissions of greenhouse gases called for by the 1997 Kyoto Protocol. For the agreement to succeed in slowing global warming, the author writes, countries should instead adopt a nationally collected tax on emissions.

Eizenstat, Stuart, "Stick with Kyoto: A Sound Start on Global Warming," *Foreign Affairs*, May/June 1998, pp. 119–121.
The author, undersecretary of State for economic, business and agricultural affairs, defends the Kyoto Protocol's emphasis on emission targets as the most promising way to reduce global reliance on fossil fuels and the threat of global warming.

Huber, Peter, "The Energy Diet That Flopped," *Forbes*, May 18, 1998, p. 306.
Efforts to save energy by making cars more fuel-efficient, insulating buildings and building energy-saving appliances and lighting have failed to reduce overall consumption, the author writes, because cheap energy has encouraged consumers to buy increasingly bigger cars, homes and appliances.

Olcott, Martha Brill, "The Caspian's False Promise," *Foreign Policy*, summer 1998, pp. 94–113.
The United States is counting on newly discovered reserves in the Caspian Sea region to help meet its oil needs, but these hopes may be dashed if the sudden influx of oil wealth destabilizes this impoverished and politically unstable region.

O'Reilly, Brian, "Transforming the Power Business," *Fortune*, Sept. 29, 1997, pp. 142–156.
The electric-utility industry is being revolutionized as states begin to dismantle the regulations that have long protected it from competition.

Reports and Studies

Chernoff, Harry, "The Impact of Industry Restructuring on Electricity Prices," *American Gas Association*, July 1998.
Average electricity prices will decline from 6.9 cents per kilowatt-hour (kWh) in 1996 to 5.9 cents by 2015, predicts Chernoff, senior economist at Science Applications International Corp., a high-technology research and engineering company. Most of the savings will benefit industrial and commercial users, however, as residential rates are expected to fall from 8.4 cents per kWh to 7.7 cents over the same period.

Flavin, Christopher, "Wind Power Set New Record in 1998: Fastest Growing Energy Source," *Worldwatch Institute*, Dec. 30, 1998.
Germany, Spain and the United States have contributed to the growing development of wind power, one of the most promising renewable-energy sources. Wind-generating capacity has doubled over the past three years.

Holt, Mark, "Transportation of Spent Nuclear Fuel," *CRS Report for Congress*, Congressional Research Service, May 29, 1998.
As Congress grapples with the issue of storing the country's mounting supply of radioactive waste from nuclear power plants, a key issue involves the safety of transporting this lethal material to a central repository.

Parker, Larry B., "Electric Utility Restructuring: Overview of Basic Policy Questions," *CRS Report for Congress*, Congressional Research Service, Jan. 28, 1997.
Lawmakers are faced with a number of complex issues in deciding how to craft federal guidelines to lend uniformity to utility deregulation.

The Next Step

Additional information from UMI's Newspaper and Periodical Abstracts™ database

Energy Policy

"DOE Again Proposes Energy Policy for U.S.," *Oil & Gas Journal*, Feb. 23, 1998, p. 34.
The Department of Energy has again proposed a policy to outline U.S. energy goals, objectives and strategies. The first goal of the draft is to improve the efficiency of the U.S. energy system.

Halbouty, Michel T., "Using All Energy Resources a Matter of National Survival," *Houston Chronicle*, Sept. 13, 1998, p. C4.
The author says that all domestic energy resources—petroleum, coal, nuclear, solar, hydropower, wind, oil shale and biomass, to name only a few—should be continuously researched for improved usage. But the article says that key U.S. resources—petroleum, coal and nuclear energy—are not being domestically produced in quantities or in manners adequate to meet future energy demands.

Falling Oil Prices

Flanigan, James, "Why Oil Prices Don't Behave the Way They Used To," *Los Angeles Times*, March 1, 1998, p. D1.
For more than 20 years, whenever tensions rose in the oil-rich Middle East, the price of crude oil, jet fuel, heating oil and gasoline also rose. But oil prices have fallen 33 percent in the last six months, from almost $23 a barrel to $15.35 today. And the decline gained momentum as a showdown with Iraq neared.

Hamilton, Martha M., "Pump Prices Drop, But Oil Profits Tank; Warm Weather, Asian Crisis Are Likely to Hurt Earnings," *The Washington Post*, Jan. 9, 1998, p. G1.
The crisis in Asia, the anticipated resumption of oil exports by Iraq and warm weather that curtailed demand for heating fuel in the United States and Europe have combined to push oil prices down to about $17 a barrel, compared with $26 this time last year.

Ibrahim, Youssef M., "Falling Oil Prices Pinch Several Producing Nations," *The New York Times*, June 23, 1998, p. A6.
The recent drop in oil prices, which began in October and accelerated recently, has prompted several oil-producing countries to take some of the medicine that high oil prices once imposed on oil-guzzling countries like the United States.

Schlesinger, Jacob M., "Falling Oil Prices Soften Inflation Woes—As Cost of Energy Plunges To the Levels of 1980s, U.S. Economy Benefits," *The Wall Street Journal*, Dec. 1, 1998, p. A2.
Yesterday's plunge in energy prices underscores a critical element in the American economy's surprising resilience of late: Falling oil costs are aiding growth, much the way the 1970s oil shocks strangled it. The biggest boon is lower inflation.

Telhami, Shibley, "The World; Middle East; Falling Oil Revenues Won't Make Gulf States Pushovers," *Los Angeles Times*, Dec. 13, 1998, p. M2.
As the Arab members of the Gulf Cooperation Council met last week to devise a strategy to halt the dramatic decline in oil prices, the United States found itself in an ironic position. Twenty-five years ago, it worried that the growing power of oil-producing nations, especially in the Middle East, posed a significant challenge to U.S. interests in the region. Today, it faces the opposite worry: that plummeting oil prices may undercut the stability of friendly Arab states.

Global Warming

"Turning Down the Heat," *America*, Dec. 12, 1998, p. 3.
An editorial discusses the significance of the signing of the Kyoto Protocol, in which the United States and other nations pledged to reduce their greenhouse emissions. The U.S. energy industry is lobbying against Senate ratification of the treaty.

Edwards, Chris, Peter Merrill, Ada Rouss and Elizabeth Wagner, "Cool Code: Federal Tax Incentives to Mitigate Global Warming," *National Tax Journal*, September 1998, pp. 465–483.
The Clinton administration's fiscal year 1999 budget marks a revival of interest in using the federal income tax code to influence energy demand. In the 1970s, Congress enacted tax incentives for energy conservation and alternative fuels. Now the focus is on tax incentives to mitigate global warming.

Oil Producers

"Business: Oil To Be Number One Again," *The Economist*, Nov. 28, 1998, p. 64.
In August, British Petroleum made the first attempt to stir up the oil industry when it announced the acquisition of Amoco for $48 billion. Exxon Corp. is in talks to acquire Mobil Corp.

Caragata, Warren, and Jennifer Hunter, "Union of Giants," *Maclean's*, Dec. 14, 1998, pp. 44–46.
The authors discuss the 1998 merger between Exxon Corp. and Mobil Corp., two of the biggest pieces of the Rockefeller oil trust, which had to be split in 1911.

Kraul, Chris, and James F. Smith, "Venezuela, Mexico Are Gaining More Clout; Energy: Oil Producers Outside Middle East are Helping to Sway the World Market," *Los Angeles Times*, March 24, 1998, p. D1.
The role of Venezuela and Mexico in driving Monday's surge in oil prices demonstrates the shifting power balance on the world petroleum scene as producers outside the Middle East, especially Venezuela, gain greater clout.

Renewable Energy Sources

"Infinite Power; Renewable Energy May Need Boost, But Must Compete," *Houston Chronicle*, Dec. 14, 1998, p. A26.
Polling data by consumer advocate groups suggest most Texans favor the use of renewable energy sources to ensure future power availability, reduce pollution and lessen dependence on foreign oil. The data also indicate that people would be willing to pay higher prices for those benefits. Wind, a resource Texas possesses in abundance over its plains, shows promise as a viable form of clean, renewable, cost-efficient energy.

Coffman, Keith, "Farmers Urged to Harness Solar, Wind Power," *The Denver Post*, Jan. 15, 1999, p. F6.
An engineer with the U.S. Department of Energy's National Renewable Energy Laboratory acknowledges that persuading farmers and ranchers to consider solar and wind power for their operations requires a new way of thinking. He and a fellow engineer are staffing the lab's information booth at the National Western Stock Show and Rodeo, and they say all industries, including agriculture, can benefit from recent advances in non-fossil fuel energy sources.

Eaton, John, "'Cool Car' Technology Could Improve Ventilation and Improve Fuel Efficiency," *The Denver Post*, July 26, 1998, p. L1.
The "Cool Car" project at the U.S. Department of Energy's National Renewable Energy Laboratory (NREL) involves testing various materials for windshields and interiors, as well as advanced air-conditioning, heating and ventilation systems. "This system can reduce fuel consumption by 5 percent and lower consumer fuel costs by at least $6 billion a year, and save nearly 7 billion gallons of gasoline annually," said Rob Farrington, NREL's project leader.

Fillion, Roger, "Sources of Power to be Disclosed; Rule Could Aid Renewable Energy," *The Denver Post*, Jan. 23, 1999, p. C1.
Colorado power customers will get to find out where their electricity originates under new state rules that environmentalists hope will unleash demand for renewable forms of energy such as wind and solar power. Under the rules, Public Service Company of Colorado and Pueblo-based West Plains Energy must itemize how coal, natural gas and other fuels account for the utilities' electricity generation and purchases.

Wolfe, Bertram, "Is There Now a Need for Nuclear Energy?" *San Francisco Chronicle*, Dec. 8, 1998, p. A27.
The author argues that population growth presents two major concerns: the future availability of energy, the lack of which could lead to war, and the potentially disastrous global warming caused by burning fossil fuels. He says people should work to establish solar power as a practical, major, energy supply.

Zonkell, Phillip, and Dennis Rodkin, "Software Helps Homes Save Energy and Get Green," *Chicago Tribune*, Jan. 24, 1999, p. 3.
The Alliance to Save Energy's software package, "The Energy Efficiency Connection," introduces a variety of energy-efficient options. After specifying a regional climate zone, users navigate through a virtual single-family house, choosing from 14 different energy-saving measures, including replacing windows, reducing air leaks and adding insulation. The Center for Renewable Energy and Sustainable Technology's SolarSizer designs and sizes custom-built, home-based photovoltaic systems, which convert sunlight into electricity. Novices to the mechanics of solar energy can use this program with no fear.

Utility Deregulation

Pionke, John, "In Electric Deregulation, Remember the Consumers, NLC warns Congress," *Nation's Cities Weekly*, Jan. 18, 1999, p. 12.
The National League of Cities is encouraging Congress to make sure that electric-utility deregulation will benefit consumers by ensuring that federal actions safeguard local service needs and enhance state laws.

Pristin, Terry, "Energy Deregulation Begins in New York," *The New York Times*, April 2, 1998, p. B10.
The deregulation of the electric-utility industry came to New York yesterday when independent power suppliers began accepting applications from customers who want to buy their power from a company other than Con Ed.

Ramirez, Charles E., "Edison Proposes Deregulation Guidelines," *Detroit News*, April 8, 1998, p. B1.
The Detroit Edison Co. is offering a new set of guidelines for the deregulation of Michigan's electric-utility industry—even though it doesn't expect the state legislature to pass critical enabling legislation this year.

The following article by Mary Cooper (Nov. 5, 1999 issue of the *CQ Researcher*) discusses the ever-growing phenomenon of suburban sprawl and its impact on the aesthetic character of landscape and on the economic health of communities. Since the end of World War II, farmland and forests have been converted to residential subdivisions and shopping malls, a trend that slowed during the economic recession of the 1980s, but has since rebounded. When farms and ranches are converted to suburbs, land prices go up, causing increases in property tax assessments and forcing more farmers and ranchers to sell their property to developers because they can no longer afford to earn their living off the land. This positive feedback loop is leading not only to economic hardship for farmers, but also to landscapes blighted by construction of roads, buildings, and parking lots. As Cooper notes, once scenic vistas have become degraded by structures in which no attempt has been made to minimize the aesthetic impact, and acres of land clearcut for development are ironically named for the landscapes they have destroyed (e.g., Woodland Acres).

Though not mentioned explicitly in Cooper's article, suburban sprawl also leads to a variety of negative geologic and biologic impacts, including increased flooding hazards, waste of natural resources, and habitat fragmentation. As more land is covered by concrete, asphalt, and buildings, rainfall that once infiltrated the soil instead runs off into stream channels, increasing the magnitude and frequency of flooding (see Press & Siever, pp. 255, 257). Suburbanites who commute to work in adjacent cities, and who often must travel long distances to get to shopping centers, schools, and doctors' offices, are completely dependent on their cars for transportation and consume vast amounts of oil (see Press & Siever, pp. 512–513, for a discussion of fossil fuel consumption; Merritts, De Wet, & Menking, pp. 334, 337–338, 357–358, 401–404). Finally, critical habitats are lost or fragmented, threatening wildlife and plant species. In response to these problems, the federal government and state and local groups are acting to protect land from development by buying large tracts and by passing legislation that prohibits construction outside of existing urban and suburban areas. Cooper notes that these efforts have been met with considerable criticism. Property rights advocates question the government's ability to manage its existing lands and also believe that the government is trying to deprive them of their rights to do as they please with their lands. At the same time, support for open space protection appears to be growing as more citizens find that suburban sprawl reduces their quality of life.

Saving Open Spaces

Are land-conservatism efforts in the public interest?

Last November, voters across the country expressed their frustration over suburban sprawl and the traffic congestion and visual blight that accompany it by approving more than 120 ballot initiatives to conserve undeveloped land. The Clinton administration and a number of lawmakers support a broader federal role in land conservation, mainly through an increase in federal acquisition of private property for parks and other public land. Property-rights activists say such efforts waste taxpayers' money and put landowners under unfair and overwhelming pressure to sell their land. Meanwhile, state and local governments are acting on their own to conserve green space with "smart-growth" initiatives to limit new development, and citizen-run, nonprofit land trusts are sprouting up all over the country to buy up open land.

The Nature Conservancy purchased the 279-acre San Miguel River preserve near Placerville, Colo., because it supports a unique plant community featuring the rare narrowleaf cottonwood, the Colorado blue spruce and the thinleaf alder. (The Nature Conservancy/Harold E. Malde)

THE ISSUES

83 • Should the federal government spend more to conserve open space?
• Should tax breaks for conservation easements be expanded?
• Are private acquisitions of land for conservation in the public interest?

BACKGROUND

90 **Patterns of Development**
The growing postwar demand for larger houses and more open space spawned a half-century of suburban growth.

90 **Buying Public Land**
The 1949 Federal Lands to Parks Program and the 1964 Land and Water Conservation Fund are key land-acquisition tools.

CURRENT SITUATION

92 **Clinton's Initiatives**
Major administration programs include the $1.2 billion Lands Legacy initiative.

92 **Action in Congress**
Congress is considering several land-conservation proposals, including the Conservation and Reinvestment Act.

93 **State, Local Initiatives**
Half the states now protect farmland.

97 **Land Trusts**
Most of the nation's 1,200 private land-conservation groups were organized in the past 15 years.

OUTLOOK

98 **Action Stalled?**
Congress is unlikely to approve major proposals affecting federal land-conservation policy this year.

SIDEBARS & GRAPHICS

84 **Most States 'Not Effective' in Protecting Open Spaces**
Sierra Club survey names 22 states.

88 **Land Trusts Boomed in Rockies and South**
But New England still has the most.

91 **Local Land Trusts Doubled Their Efforts**
Nearly 5 million acres were under protection in 1998.

94 **How Crested Butte Manages Growth**
A land trust buys open space in the Colorado ski resort.

96 **Efforts to Save Evans Farm Came Too Late**
The popular spot near Washington, D.C., will become a housing development.

101 **Chronology**
Key events since 1803.

102 **At Issue**
Should the federal government buy more land to help conserve open space?

FOR FURTHER RESEARCH

103 **Bibliography**
Selected sources used.

104 **The Next Step**
Additional articles from current periodicals.

Note: For more information on this topic, please see the following pages in Press and Siever's *Understanding Earth*, Third Edition: pp. 255, 257; pp. 512–513, discussion of fossil fuel consumption; and in Merrits, De Wet, and Menking's *Environmental Geology:* pp. 334, 337–338, 357–358, 401–404.

Saving Open Spaces

By Mary H. Cooper

The Issues

First, a "For Sale" sign crops up in a field or woodland just beyond the new mall outside town. Soon the bulldozers arrive to level the landscape, including most of the trees, and to slash a road through the property. Then new houses erupt from the bare ground, and another sign pops up. "Welcome to Woodland Acres," it announces.

To developers and new residents, new suburban subdivisions embody nothing less than the American Dream, a chance to trade cramped apartments in chaotic, noisy cities for the good life of clean air, quiet surroundings and home ownership.

"Americans want open space in their back yards and beyond," says Neil Gaffney, director of environmental communications for the National Association of Home Builders. "And the best place for that is the suburbs."

But to others, so-called suburban sprawl and its tidy subdivisions are rapidly chewing up one of America's most cherished assets—open space.

"You see residential subdivisions spreading like inkblots, obliterating forests and farms in their relentless march across the landscape," noted Richard Moe, president of the National Trust for Historic Preservation. "You see a lot of activity, but not much life. You see the graveyard of livability." [1]

However the scene is viewed—nightmare or American Dream—it is familiar to communities across the United States. In the East, the megalopolis stretching from Boston to Washington continues to spread, consuming the few remaining tracts of open land. Rapid growth around Phoenix, Ariz., and other cities of the desert Southwest is threatening the region's water supplies. Even picturesque, rural communities in the Rocky Mountains are mourning the loss of the open spaces that drew many residents there in the first place. [2]

"Open space is disappearing at an alarming rate," says Russ Shay, director of public policy for the Land Trust Alliance, a Washington-based organization of individuals and groups working to conserve land. "People are concerned about the lands right around their communities, where rapid development is not only destroying landscapes but also contributing to traffic congestion, leaving them hours instead of minutes away from their destinations."

Many Americans seem to agree. Last year, voters approved 168 of the 240 state and local ballot initiatives that provided more than $7.5 billion in new funding for open space preservation. [3] Some of the measures created boundaries that halted development beyond already-developed areas; others raised local taxes to buy land and keep it out of developers' hands.

Concern over disappearing open space is also reflected by the proliferation of land trusts in communities across the country. Typically, these private conservation groups buy land outright or pay a landowner to refrain from developing it. The best-known trust is the million-member Nature Conservancy, based in Arlington, Va., an international organization that has helped protect more than 58 million acres around the world.

Most land trusts, however, are small, community groups. In the past decade, the number of local and regional trusts has grown more than 150 percent, to some 1,200. During that period, local trusts have permanently protected 2.7 million acres from development. [4]

Although support for environmental-protection laws often pits Democrats against Republicans, land conservation is one of the few issues that defy partisanship. In fact, the Democratic governor of Maryland and New Jersey's Republican chief executive backed two of the most sweeping state open-space initiatives ever proposed. In Maryland,

Most States 'Not Effective' in Protecting Open Spaces

Only 22 states are effective in preventing the loss of open space, according to a recent survey by the Sierra Club. In the most effective states, parks and open space are purchased outright. Moderately effective states generally have passed initiatives to hold lands in trust.

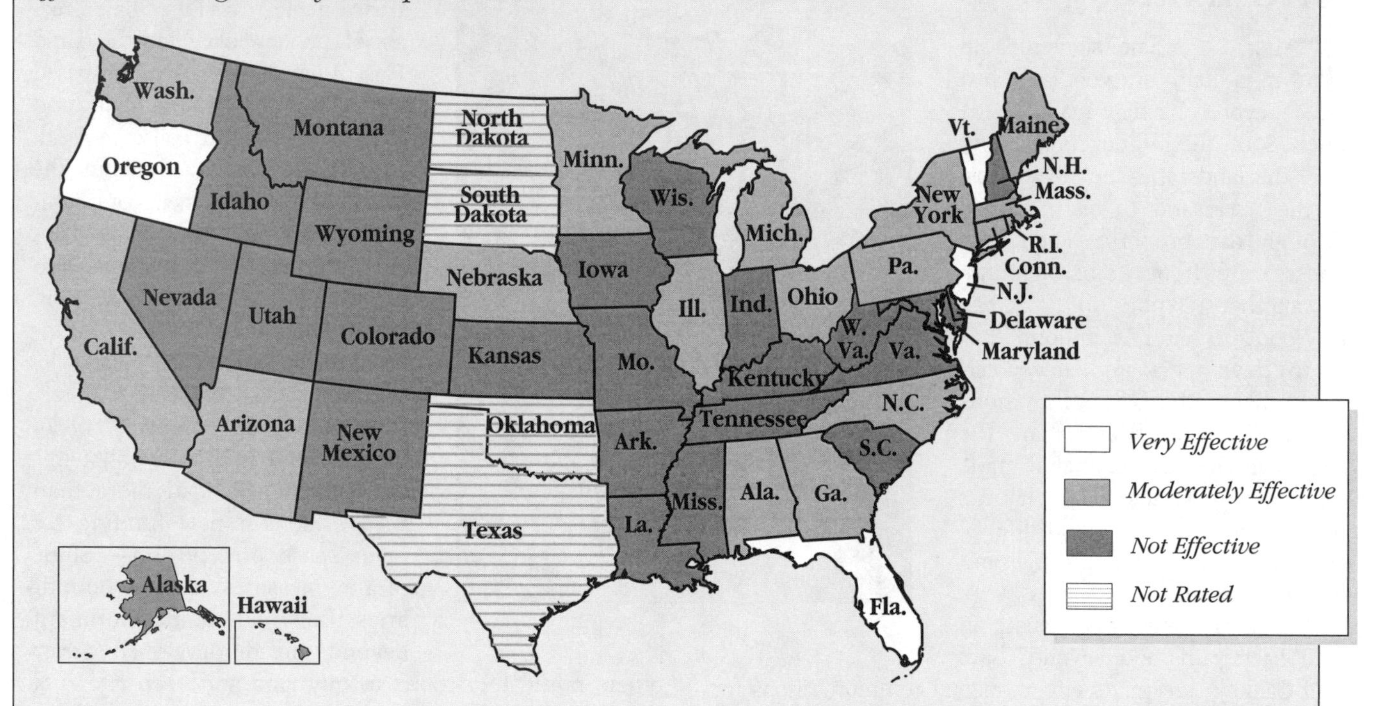

Source: "1999 Sierra Club Sprawl Report." The rankings were based on information from the National Governors' Asssociation, private land trusts and Sierra Club grass-roots activists.

Gov. Parris N. Glendening's "smart growth" initiative has barred much new development outside urban areas. Gov. Christine Todd Whitman of New Jersey has launched an ambitious plan to set aside 1 million acres in her densely developed state. Voters last year approved her plan to spend $1.9 billion on the project.

Although land-use policy has traditionally fallen under state and local government jurisdiction, the Clinton administration is calling for new federal initiatives to bolster the inventory of federally owned open land and to encourage land conservation at the state and local levels as well. In January, President Clinton announced a Lands Legacy initiative, which called for almost $1 billion in federal money to protect open lands as well as battlefields, archaeological sites and historic structures. In late October, however, lawmakers gutted the initiative by providing far less funding for land acquisition than the president had requested (*see p. 98*).

Undaunted by congressional reluctance to endorse his open-space initiative, Clinton announced on Oct. 14 one of the most sweeping land conservation proposals in U.S. history, calling for up to 40 million acres of national forest to be protected from development. [5] In addition to direct federal intervention to save open space, the administration would provide greater federal support for local land-conservation efforts. In January, Vice President Al Gore called for $700 million in tax credits to help state and local governments build more "livable" communities.

"Many of the green places and open spaces that need protecting most today are in our own neighborhoods," said Gore in announcing the administration's Livability Agenda. "In too many places, the beauty of local vistas has been degraded by decades of ill-planned and ill-coordinated development. In too many places, people move out to the suburbs to make their lives, only to find they are playing leapfrog with bulldozers." [6]

The Nature Conservancy's San Pedro River Preserve in Arizona features cottonwood-willow forests, the rarest forest type in North America. The property and surrounding lands support 350 species of migratory and resident birds.

The Nature Conservancy/Harold E. Malde

Critics of federal and state efforts to curb sprawl say they are not only redundant but may backfire as Americans discover that government regulations inhibit their freedom to move where they want.

"We have worked under land-use regulations for decades, so I'm befuddled, to say the least, to hear that Gore or Glendening invented the concept of bettering communities through planning," says John T. "Til" Hazel, a lawyer and leading land developer in fast-growing Fairfax County, Va. "In any case, who is going to decide what land is to be set aside; who has the right to decide what the habits of our society are going to be?"

A number of initiatives to preserve open space are now before Congress. Every year since 1964, Congress has had the authority under the Land and Water Conservation Act to spend up to $900 million a year on open space preservation. But for most of that time there was a deficit in the federal budget, prompting lawmakers to use most of the money for other programs. Now that there is a budget surplus, lawmakers are more willing to use the Land and Water Conservation Fund for its original purpose, and bipartisan proposals requiring full funding of the authorized amount are now before Congress.

"Lawmakers are listening to people's complaints about how long their commutes are and how they don't like to see the spaces they loved as children destroyed," says Helen Hooper, director of congressional affairs at The Nature Conservancy. "They have seen how people are raising their own taxes to save places in their communities and understand they can now get away with spending money on saving land."

Property-rights advocates oppose permanent and full funding of the Land and Water Conservation Fund because they fear it would escalate what they see as a massive land grab by the federal government in which landowners are coerced into selling their property. [7]

"We're not opposed to some acquisition of private land," says Chuck Cushman, executive director of the American Land Rights Association in Battle Ground, Wash. "But these proposals would really mean massive land acquisition, enormous abuses and nothing but trouble for landowners."

Critics of the property-rights argument against expanding public lands respond that taxpayers also have rights—rights to uncluttered landscapes, undisturbed wetlands and intact forests where they can escape the chaos of modern life. "It's easy to focus on property-rights issues," says Ralph Grossi, president of the American Farmland Trust. "But the flip side of property rights is public rights."

Preserving open space promises to be a leading issue in the upcoming presidential election campaign. Alone among the major contenders, Gore has placed land conservation near the top of his policy agenda. His chief rival for the Democratic nomination, former Sen. Bill Bradley of New Jersey, has limited his comments on the issue to saying the vice president has failed to live up to his promise to protect the environment. For his part, Texas Gov. George W. Bush, the leading Republican contender for the presidency, strongly advocates property rights and opposes federal intervention in state and local land-use issues.

As the debate over open space and property rights heats up, these are some of the questions being asked:

Should the federal government spend more money to conserve open space?

Thanks to efforts pioneered by President Theodore Roosevelt and others, the federal government today owns about 563 million acres, or about a quarter of the 2.3 billion acres that comprise the 50 states. Although almost a third of the federally owned land is in Alaska, the government also owns more than half the land in Idaho, Nevada, Oregon and Utah. [8] Although much of this property is open to cattle grazing, timber cutting and mining, about 104 million acres are designated as wilderness areas, where development and access by motorized vehicles are banned. [9]

As public concern over the loss of open space has mounted in recent years, the Clinton administration has undertaken some of the most sweeping land acquisitions in decades. In September 1996 the president designated 1.9 million acres of public land in southern Utah as the Grand Staircase-Escalante National Monument. In August of this year, the government agreed, pending congressional approval, to buy the 95,000-acre Baca Ranch in New Mexico's Jemez Mountains for $101 million and turn it into a park. The property is home to one of the country's largest elk herds as well as bears, mountain lions and eagles. In September, federal officials completed a $13 million purchase of 12,000 acres of ranchland abutting Yellowstone National Park, also to be preserved as permanent open space. [10]

For fiscal 2000, which began Oct. 1, the president intensified his land-acquisition efforts. He proposed to spend $440 million for land purchases—up from $328 million in fiscal 1999—as part of his $1 billion Lands Legacy initiative, which lawmakers recently slashed. He also asked Congress to pass a law to ensure that the Land and Water Conservation Fund receive its full annual funding of $900 million, which would be used to protect open space.

Conservationists strongly support increasing spending to augment the federal government's landholdings. "The promise of the Land and Water Conservation Fund has never been fully kept," Interior Secretary Bruce Babbitt said in September. "It was set up to provide almost $1 billion a year through direct federal purchase and through grants to states that could then be used to protect land and open space. But rather than have these funds flow automatically, Congress has required that it appropriate the money every year. Unfortunately, Congress has usually approved less than one-third of the money available." [11]

According to a recent opinion poll, the public supports federal purchase of open space. When asked how the Land and Water Conservation Fund should be used, more than 80 percent of the respondents said the money should go toward expanding existing national parks, forests and recreation areas or creating new ones, rather than reducing the deficit. [12] The concern appears to extend beyond the suburban landscape to include remote areas as well.

Another survey asked, "How would you rather see $2 billion of the surplus spent—to buy and protect wildlands and other natural places or increase military defense spending? Half the respondents picked spending on the environment; 34 percent chose defense spending. [13]

"The best way to ensure that these special places remain special for our children and grandchildren," states a Sierra Club report, "is to acquire the land and hold it in public ownership for public purposes." [14]

Property-rights advocates say that federal acquisition of private land often violates the rights of landowners, who come under pressure to sell their property from heavy-handed government bureaucrats. "Experience has demonstrated that even the most reluctant landowner can be converted into a 'willing seller' by a persistent and determined agency," wrote Stan Leaphart, executive director of the Citizens' Advisory Commission on Federal Areas in Alaska, a state agency in Fairbanks. "We would submit that for each property owner who freely offers their property for sale to a federal land-management agency, there are dozens who are eventually manipulated into a situation where they are left with no real choice but to sell." [15]

Property-rights advocates also predict that federal acquisition of additional land will hurt landowners everywhere. "It's a billion-dollar boondoggle, the ultimate land grab that keeps on grabbing," said Cushman of the American Land Rights Association in a recent e-mail alert. "Do we really want the federal government buying up even more private land and taking it off the tax rolls? The federal government already owns nearly a third of the entire country. The purchase of more land will cause your property taxes to increase on the remaining private lands." [16]

Critics of federal land policy also say the government cannot properly maintain the land it already owns. [17] Rep. Ralph Regula, R-Ohio, chairman of the Interior Appropriations Subcommittee, for example, opposes Clinton's proposal to spend $747 million for new land purchases.

"The president's Lands Legacy initiative flies in the face of my panel's No. 1 priority for the past four years—focusing limited resources on addressing the critical backlog of maintenance problems and operational shortfalls in the national parks, wildlife refuges, national forests and other public lands, which now exceed $12 billion," Regula said. "It is simply irresponsible to take on new land responsibilities and give grants to cities, states and private institutions when we cannot afford to adequately take care of our primary federal responsibilities—the public lands." [18]

Some conservationists agree that more funding should go toward maintenance. "We need to address the crisis in our national park system, which is the real patrimony of our nation," says Paul C. Pritchard, president of the National Park Trust, which raises money to buy private property within or next to federal lands to keep them from being developed. "You can't just say we need more land; we also need more people to manage it. We aren't managing properly the land we have now, but the government doesn't want to address this, and the public is losing out."

But Pritchard also advocates more funding to purchase private land, especially "inholdings"—private property that lies within federally owned land. His organization has identified 200,000 acres of national park inholdings that are in "imminent" danger of being developed or resold and asks Congress to come up with the $70 million the trust estimates it would cost to buy them for the public good. [19]

"We can't save all the valuable parkland," says Pritchard. "The boat will go; we're just trying to keep the lifeboats from sinking by plugging the holes."

Should tax breaks for conservation easements be expanded?

Many advocates of land conservation are wary of federal involvement in conservation efforts that go beyond outright acquisition of land for parks, monuments and other federal holdings. Land-use decisions and zoning issues traditionally have fallen to state and local governments, and some observers fear that federal funding of open-space initiatives opens the door to federal intrusion into local decision-making.

"I believe that fighting sprawl, planning or growth and preserving open space are primarily issues for state and local governments," Rep. Joseph M. Hoeffel, D-Pa., told a congressional committee. "States and municipalities must deal with these issues, develop their own plans and make local decisions about what should be saved." [20]

Federal incentives are another matter, however, especially when they take the form of income tax breaks to encourage conservation. "We need national vision with local control to fight urban and suburban sprawl," Hoeffel said. "I believe it is important for the Congress to establish programs that provide incentives for preservation and conservation goals to be achieved. It is appropriate for the federal tax system to reward those who choose to preserve our natural resources and protect our environment."

The tax code already offers landowners an incentive to conserve land by allowing them to deduct a percentage of the value of their property from their income taxes. Some conservation advocates say the tax code should offer bigger rewards for saving open space.

"There are obviously a variety of causes for this move away from our cities," said Rep. Nancy L. Johnson, R.-Conn. "But I do not believe that our tax code should be one of those causes." [21] Earlier this year, she introduced a measure that would increase the current income tax deduction for land donations from 30 percent to 50 percent of the value of the land and extend the time for taking the deduction from six years to 20 years.

The 1997 Taxpayer Relief Act, the first new tax incentive for land conservation in more than a decade, also allows executors to reduce a landowner's estate tax liability by donating a conservation easement on a property. The easement allows the land to be farmed or logged, but it can never be subdivided for development purposes, even after it changes hands.

In order to qualify for such a tax break, however, the property must be near an urban area or existing parkland.* That provision excludes much of the country, including most of the Midwest. Johnson's bill, as well as a proposal by Sen. Max Baucus, D-Mont., would eliminate the geographic requirements.

"Through some very simple tax measures," Johnson said, "we can give back a little in the form of a tax deduction to those who sacrifice high profits that come from selling their land to those who instead donate for the good of us all."

Some critics say tax incentives to conserve land amount to nothing more than tax breaks for the wealthy, who already benefit from numerous loopholes in the tax code.

"In short, the conservation easement gizmo uniquely satisfies not one but two compelling needs of premillennial Homo Liberalus Americanus," says an editorial in *The Wall Street Journal*: "The need to reduce his tax burden and the need to make a show of doing something for the environment."

The editorial cites the extensive use of easements by celebrity landowners on Martha's Vineyard, Mass., and in wealthy enclaves of the West, such as Jackson Hole, Wyo., and Big Sur, Calif. Instead of reducing estate taxes for individuals who donate easements, the editorial proposes, "far better would be simply to repeal the whole tax, which would allow regular families the tax benefits now effectively confined to those with, say, a beachfront mansion or mountain." [22]

Conservation supporters reject the editorial's argument as totally distorted. "We don't deal with a lot of wealthy people, just farmers and ranchers," says Chris Montague, president of Montana Land Reliance, a statewide, private land trust. It has used donated conservation easements to protect some 322,000 acres of private land in Montana—about a fifth of all the land protected by private state and local land trusts in the United States. "These people really are giving up something huge when they donate a conservation easement—the right to subdivide their land—and they should be

*The 1997 American Farm and Ranch Protection Act allows the exclusion from taxable estates of 40 percent of the value of open land subject to a qualified conservation easement. The exclusion is capped at $200,000 and applies only to lands within 25 miles of a Metropolitan Statistical Area or a national park or federal wilderness area, or within 10 miles of a National Urban Forest.

Land Trusts Boomed in Rockies and South

The number of nonprofit state, local and regional trusts more than doubled from 1988 to 1998 in four regions, led by the Rocky Mountain states. New England, which gave birth to the land-conservation movement, still has the largest number of land trusts. Five states with no trusts in 1988 — Arkansas, Hawaii, Mississippi, Nevada and Utah — today each have at least one.

U.S. Land Trusts	1988	1998
Rocky Mountains (Colo., Idaho, Mont., Utah, Wyo.)	20	52
Southwest (Ariz., N.M., Okla., Texas)	15	37
South (Ala., Ark., Fla., Ga., Ky., La., Miss., N.C., S.C., Tenn., Va., W.Va.)	65	142
West (Alaska, Calif., Hawaii, Nev., Ore., Wash.)	83	173
Mid-Atlantic (Del., D.C., Md., N.J., N.Y., Pa.)	117	222
Great Lakes (Ill., Ind., Mich., Ohio, Wis.)	84	145
New England (Conn., Mass., Maine, N.H., R.I., Vt.)	336	417
Plains (Iowa, Kan., Minn., Mo., N.D., S.D., Neb.)	21	23
Total	741	1,211

Photo: Lower Day Mountain, Acadia National Park (Maine Coast Heritage Trust)

Source: Land Trust Alliance, Oct. 1, 1998

rewarded for it," he says. "It's a very cost-effective way to achieve land conservation."

Some developers are of two minds about tax incentives to save open space. "When we first looked at tax incentives for conservation easements, they looked like just another land grab," says Gaffney of the Home Builders Association. "But then we realized that parkland is a positive addition to a community."

The association estimates that up to 1.5 million new housing units will have to be built each year over the coming decade to accommodate an expected 30 million increase in the U.S. population.

"The crucial point in our view is that open space conservation must be decided at the local level and be part of a comprehensive plan that allows for new housing," Gaffney says.

Open Lands, a Colorado land trust, purchased the 3,200-acre Evans Ranch, near Denver, to block a 1,100-home residential development. The property now supports five working ranches.

Land Trust Alliance

Are private acquisitions of land for conservation in the public interest?

Mounting concern over sprawl in recent years has prompted the creation of more than 1,200 regional, state and local land trusts across the country. "The reason why land trusts have enjoyed fantastic growth," says Pritchard of the National Park Trust, "is that preservation of open space is one of the great American values."

Some property-rights advocates have no problem with local land trusts, which they say reflect the desires of the communities where they operate. "A lot of the local land trusts are really admirable because the locals get together and decide they want to save some land," says Myron Ebell, a property-rights lobbyist and environmental policy analyst at the Competitive Enterprise Institute. "They are an example of the voluntary associations of people accomplishing things through private and local means that [Alexis] de Tocqueville described as the genius of America. But the big national trusts are truly evil." [23]

Ebell targets his heaviest criticism at The Nature Conservancy, by far the largest land trust. This million-member organization has helped protect 11.3 million acres nationwide since it was founded in 1951 to protect habitat for plants and animals. Despite the conservancy's status as a private, nonprofit organization, Ebell says that it often buys private property only to turn around and sell it at a profit to the government. "Like the other big trusts, The Nature Conservancy is just a halfway house to federal ownership of private land," he says. "It's a profit-making enterprise; it keeps getting bigger and bigger because it makes more and more money off these deals."

Not so, Hooper says. "We have a policy that we do not make a profit when we step in as a middleman. In fact, as a matter of practice, we lose money every year on these deals."

Property-rights supporters also maintain that land conservation efforts often spell disaster for local economies. "When they buy a ranch and shut it down, they disrupt whole communities in rural America," Cushman says. "A ranch isn't just some guy with a pickup truck; it may be all that's holding a small town together. Taking that ranch out of production takes taxes out of the system. Damn few of these deals are good for the government, and they certainly aren't good for the country."

But land trust advocates say opposition to their goals is waning fast. "Support for land acquisition for the pubic good is now just as strong in the West, even the Intermountain West, as it is in the Northeast, where every acre counts," says Shay of the Land Trust Alliance. "The Nature Conservancy is the biggest land trust in America, and it's fun to pick on winners," he says. "But the conservancy would not be successful if it violated the will of local communities. Indeed, the conservancy's success is due to the fact that it has extensive local roots and maintains chapters in every state."

Whatever the merits of national or local land trusts, conservationists point out that objections to protecting open space are inconsistent with the critics' overall property-rights agenda.

"The property-rights folks can't have it both ways," says Grossi of the American Farmland Trust. "If they want to protect the rights of landowners, they have to include among those rights the right to sell the right to develop or subdivide their own property."

Background

Patterns of Development

For much of the 20th century, population growth in the United States targeted urban areas, where the promise of industrial jobs in cities lured millions from small towns and farms. Beginning with the economic boom that followed World War II, however, housing development spread outside city centers in response to growing demand for larger houses, cleaner air and more open space. Suburban development continued to spread over the next 50 years, eventually producing the "sprawl" that has given rise to today's land conservation movement. [24]

Since the early 1970s, when the declining industrial centers of the Midwest's Rust Belt and other congested cities began losing population to the suburbs, concern over the loss of open space has also emerged in many towns and rural areas. After a lull in the 1980s, when a protracted recession hit farmers and rural industries and halted the trend, the exodus from cities resumed.

Today, many non-urban areas are experiencing a "rural rebound," as retirees, disenchanted city-dwellers and computer-equipped "telecommuters" come in search of a better life. [25] In many cases, the growing demand for housing inflates land values, prompting farmers and ranchers to subdivide vast tracts and raising concerns among residents that the open space that is part of their rural heritage is fast disappearing. (*See story, pp. 94–95.*)

Sprawl has become an especially sensitive issue in both urban and rural areas of the South and West, where more than three-quarters of older Americans are settling in retirement. In part because of retirement trends, Las Vegas and Atlanta are among the fastest-growing cities in the country, while smaller cities and rural communities in the Rocky Mountain states are experiencing similar rates of growth. [26]

At the same time that privately owned open space has been consumed for housing and industry, public land also has come under pressure from development. According to the National Park Trust, many state and national parks and forests are threatened by the increasing number of "inholdings," or private property within public land. The organization reports that inholdings have grown by 1.6 million acres, or 35 percent, over the past decade, bringing the total to 84 million acres.

"The danger that this land could be sold for development, bulldozing, clear-cutting or for other destructive purposes constitutes the single greatest threat to the system of national and state parks," Pritchard said. [27] New inholdings typically are created when public land expansions envelop private property.

The Nature Conservancy/Harold E. Malde

Recreationists enjoy Long Pond in the 6,000-acre Pocono Preserve in Pennsylvania. The Nature Conservancy property supports one of the most significant moth and butterfly habitats in the state.

As support for land conservation mounts, some critics say public reaction to sprawl is overblown. The San Francisco-based Pacific Research Institute for Public Policy, which promotes a limited role for government, estimates that development consumes only .0006 percent of the land in the continental United States each year and that the rate of sprawl is lower today than in the 1950s or '60s. [28]

Buying Public Land

The federal government has been buying land since the nation's origins. During the period of frontier expansion that ended in the late 1800s, the government spent $85 million to buy 1.8 billion acres—almost 80 percent of the country today. Thomas Jefferson made the biggest land deal when he paid France $23 million for the Louisiana Purchase in 1803, thus adding 530 million acres to U.S. territory, almost a quarter of total U.S. land today. Later acquisitions included the 1867 purchase of Alaska—378 million acres—from Russia for $7 million and the 339-million acre Mexican Cession of 1848 for $16 million.

Most of the federally acquired public land was later granted to states as they joined the Union, granted or sold to homesteaders under the 1862 Homestead Act [29] or sold to railroad companies, timber companies and other private interests.

During the 20th century, the federal government has resumed land acquisitions, though on a far less sweeping scale than in the nation's early years. Beginning with the

Local Land Trusts Doubled Their Efforts

Nonprofit local and regional land trusts had nearly 5 million acres under protection in 1998, more than twice as many as in 1988. The protected land—covering more area than Connecticut and Rhode Island—includes family farms and ranches, trails and scenic views, wetlands and forests. Three-quarters of the land is used for public recreation.

METHOD OF PROTECTION	ACRES PROTECTED 1988	ACRES PROTECTED 1998
Conservation easements	0.3 million	1.4 million
Owned by land trusts	0.3 million	0.8 million
Transfer of land to government agency or other means*	1.4 million	2.5 million
TOTAL	2.0 million	4.7 million

** Includes lands protected by trusts holding deed restrictions or negotiating for acquisition by other organizations or agencies.*

Source: Land Trust Alliance, Oct. 1, 1998

2.2-million-acre Yellowstone National Park, created by Congress in 1872, the federal government began setting aside public land and buying or exchanging private property as parks, national forests and wildlife preserves.[30] States followed suit with purchases and set-asides for state parks, forests and recreation areas.

Congress has created several mechanisms for adding to the inventory of public lands. The 1949 Federal Lands to Parks program, for example, allows for the no-cost transfer of federal lands to state, regional and local governments for use in perpetuity as parks and conservation areas. Recent military base closings have enabled communities across the country to open to the general public open land formerly reserved for the military.

Arguably the most valuable land-acquisition tool at the federal government's disposal is the 1964 Land and Water Conservation Fund Act. Created specifically to protect outdoor recreational resources, the fund provides money for federal acquisition of land and easements and matching funds for states to be used to buy and develop outdoor recreation facilities. Since it went into effect on Jan. 1, 1965, the fund has been used to create almost 7 million acres of parkland, water resources and open space as well as more than 37,000 state and local park and recreational projects.

The law authorizes Congress to appropriate up to $900 million annually for acquisitions and state grants. The money comes primarily from royalties paid by oil and gas companies for the right to drill on the outer continental shelf, which is federal property. In this way, money derived from exploiting non-renewable energy resources would be used to purchase another non-renewable resource—undeveloped land—for the public good.

But Congress has rarely appropriated the full amount authorized under the law, instead using the bulk of the money intended to protect open space to reduce the federal budget deficit. Since fiscal 1995, the state matching fund portion of the Land and Water Conservation Fund has received no funding for new grants.

Current Situation

Clinton's Initiatives

Two recent developments have refocused attention on the long-neglected fund. First, voters last November approved more than 120 ballot initiatives aimed at preserving open space and curtailing sprawl, sending a clear message to lawmakers facing re-election next year that land conservation is a high priority. Second, the federal budget surplus has freed lawmakers from the need to siphon off money from the fund to reduce the deficit. As a result, a new, largely bipartisan consensus has emerged this year between the Clinton administration and members of Congress to utilize the full Land and Water Conservation Fund as the linchpin of renewed land conservation efforts.

The administration has proposed the most ambitious set of federal initiatives to promote land conservation in decades, including a proposal to fully appropriate money for the fund. The Lands Legacy initiative, a $1.2 billion measure included in the fiscal 2000 budget request, included $440 million for the fund to be used for federal land acquisitions and $150 million for state matching grants for conservation uses. The initiative also included $200 million for coastal protection; $80 million for habitat conservation; $50 million in competitive grants for smart growth planning; $50 million for the Forest Legacy program, begun in 1997 to purchase development rights from owners of timberland; and $50 million for farmland protection.

In January, Vice President Gore unveiled the administration's $1 billion Livable Communities initiative, which would enable states and localities to raise a total of $9.5 billion through bond sales aimed at helping communities preserve green space and improve the local environment. It would be among the largest federal programs ever undertaken to curb sprawl. The proposal includes $700 million in tax credits for state and local governments, with the rest of the money to be invested in public transit and other measures to ease traffic congestion as well as steps to promote regional planning among neighboring communities.

In March, the administration and the state of California agreed to pay Pacific Lumber Co. $480 million for 10,000 acres in Northern California that includes a rare grove of ancient redwoods known as the Headwaters Forest. The tract has been set aside as a nature preserve with public access. On Oct. 14, Clinton announced plans to ban road building and logging on 40 million acres of national forest wilderness. Scheduled to take effect late next year after a period of public comment, the plan would effectively preserve a fifth of the 192-million-acre national forest system and constitutes the most sweeping land-conservation initiative since Theodore Roosevelt's administration. [31]

The administration's record on land conservation is a matter of some debate among environmental advocates. "We applaud the renewed attention the administration is giving to the issue, and we support the full funding of the Land and Water Conservation Fund as originally intended," says the American Farmland Trust's Grossi. "But we feel that the administration's proposals are heavily weighted in favor of federal acquisition of land, and not enough for the states."

Grossi would like to see more incentives for farmers to use conservation easements to protect farmland. "The federal government could play a major role without meddling in local affairs by providing the funding while leaving the decision about which land to protect to the local communities," he says.

But some environmentalists are disappointed in the administration's level of commitment to land conservation. "The Clinton administration has talked a lot, but done very little," says Pritchard, who argues that the Lands Legacy initiative does little to solve the problems of low funding for national parks. "The reason we're seeing a lot of people disenchanted with the federal government is its lack of commitment to protecting open space."

Action in Congress

Several land-conservation proposals are under consideration in Congress. In February, House Resources Committee Chairman Don Young, R-Alaska, and Rep. John D. Dingell, D-Mich., ranking member of the House Energy and Commerce Committee, introduced the Conservation and Reinvestment Act, which would permanently fund the Land and Water Conservation Fund.

The Young-Dingell bill would place some restrictions on the fund's federal program, however. It would:

- limit the fund to buying inholdings within existing federal property;
- require that two-thirds of the money be spent east of the Rockies, where federal property is relatively scarce;
- require congressional approval for purchases of more than $1 million; and
- ban the use of condemnation, the process by which the government can force landowners to cede their property.

The bill also allocates $1.24 billion to coastal states for coastal restoration and mitigation based on a formula that ties funding to the proximity of offshore oil and gas wells, a feature that environmentalists charge would encourage new drilling.

Land Trust Alliance

By donating conservation easements on their property, ranchers and farmers can continue to use the land while protecting it from development in perpetuity.

The Senate version of the Conservation and Reinvestment Act, introduced by Energy and Natural Resources Committee Chairman Frank H. Murkowski, R-Alaska, and Sen. Mary L. Landrieu, D-La., is similar to the Young-Dingell proposal. It would dedicate half the offshore drilling revenues to support a permanently funded Land and Water Conservation Fund and allocate $340 million to each of the fund's two components. It contains the same restrictions, except for a more generous ($5 million) limit on the maximum value of acquisitions that could be made without congressional approval. It also includes a similar formula for allocating coastal impact aid.

A Democratic alternative to the two bipartisan bills is the Permanent Protection for America's Resources 2000, introduced in February by California Rep. George Miller, ranking Democrat on the House Resources Committee. The $2.3-billion bill would provide full and permanent funding of the Land and Water Conservation Fund from offshore drilling revenues and other conservation programs. The fund's federal and state-matching components would each receive $450 million. The bill contains no new restrictions on federal land purchases and limits coastal impact funding to current leases, thus eliminating the incentives for new drilling present in the other two bills. In addition, the bill would create conservation easement programs for ranch, farm and timberland protection. The proposal has won the support of House Minority Leader Richard A. Gephardt, D-Mo., who last year introduced his own bill to permanently fund the Land and Water Conservation Fund.

Sen. Dianne Feinstein, D-Calif., introduced the Public Land and Recreation Investment Act, a more limited bill that would ensure permanent funding of the Land and Water Conservation Fund with no restrictions on federal acquisitions.

State, Local Initiatives

States typically set aside land for public parks, forests and recreation areas, while localities control land use primarily through zoning regulations. But in recent years, some state governments have decided to exercise greater authority over zoning issues by creating statewide development plans aimed at combating sprawl. According to a recent report by the Sierra Club, half the states have taken steps to protect farmland, mostly by compensating landowners for giving up development rights by placing conservation easements on their land. Eleven states also have passed growth-management laws barring development outside established boundaries around existing urban areas. [32]

Oregon set the stage for state land conservation efforts in 1973 when the state legislature required all cities to contain most development within clearly defined urban-growth boundaries. The requirement has attracted nationwide attention to Portland's efforts to curb sprawl. More recently, Maryland and New Jersey have taken the lead in open-space conservation. Maryland's smart-growth plan, adopted in 1997, emulates Oregon's plan. On Oct. 8, Gov. Glendening approved an additional $25 million to buy more land around the state. Gov. Whitman of New Jersey has promised to follow through on her proposal to preserve a million acres of open space and push several smart-growth restrictions before term limits force her to step down on Jan. 1, 2002. [33] In August, Gov. Roy Barnes, D-Ga., announced plans to require fast-growing counties, such as those around Atlanta, to set aside at least 20 percent of their undeveloped land for open space.

Local governments also are assuming a larger role in saving open space. Several cities in California's sprawling

How Crested Butte, Colo., Tries to Manage Growth

In the Rocky Mountain states, many 19th-century mining towns now rely heavily for survival on the booming tourist trade. Each year millions of visitors flock to the region to admire its towering mountains, aspen-studded valleys and vast alpine forests.

But the same natural resources also draw new residents, many of whom have the same idea: Buy a big hunk of land—say 35 acres—outside of town and build their dream house. As houses spring up on previously undeveloped land, they mar the pristine landscape and detract from the beauty of what draws people there in the first place.

Many popular mountain resorts in Colorado, such as Aspen, Vail and Telluride, were dramatically altered by rapid residential development in the 1980s and '90s. Housing has spread up mountainsides and for miles down valleys where cattle once grazed. The demand for housing has driven property values so high that it has created a shortage of affordable housing for local workers, who are forced to commute long distances.

To avoid a similar fate, or at least slow the inevitable growth, residents of Crested Butte, another Colorado ski town, have taken steps to preserve the

To preserve mountain vistas, the land trust in Butte, Colo., tries to consolidate new construction in developed areas.

Central Valley are considering imposing impact fees on developers of new subdivisions to slow growth and help pay for the infrastructure and schools needed to serve new communities. [34] Planners in Montgomery County, Md., a fast-growing Washington suburb, recently proposed spending $100 million over the next decade to preserve undeveloped land and historic sites. The funding, to be financed by a special bond issue, would be in addition to the $3.5 million the county receives each year for land conservation from state coffers. [35] Voters in a number of localities, including Adams County, Colo., next to Denver, Springfield, N.J., and Phoenix, Ariz., the country's fastest-growing city, are being asked to approve new growth reforms and open-space proposals in this fall's elections.

In some areas, local conservation efforts are highly controversial. When King County, Wash., bought a 12-mile stretch of railroad right-of-way to build a pedestrian and community's open space. "Aspen never made any pretense about being about anything but a highly ostentatious display of wealth, and Vail basically sold itself out," says Norm Bardeen, a Crested Butte builder. "Telluride once had a lot of charm, but you couldn't pay me to live there now. They started a day late and a dollar short by doing nothing to curb growth until after the most serious damage had been done."

Crested Butte residents became alarmed that their community was headed in the same direction in the early 1990s, when the Colorado Fuel and Iron Co. sold a forested, 2,200-acre parcel just outside of town to developers. The developers carved the land into homesites of about 35 acres each.

"The town wanted to purchase the property, but a private individual beat us in making the deal," says Jim Starr, a Gunnison County commissioner. To save the remaining open space around the town and near the Mt. Crested Butte ski resort, Starr, Bardeen and two other residents created the Crested Butte Land Trust in 1991.

The 500-member group raises money from residents and visitors alike to buy and receive donations of open space and conservation easements, which allow owners to keep the land but sell or donate their rights to develop it. The land trust also receives around $200,000 a year from a local real estate transfer tax, a 1.5 percent levy on each property sale. For the past several years, local businesses have helped support the effort by asking for a voluntary, 1 percent "sales tax" on goods and services, with proceeds earmarked for the land trust.

Cattle ranchers, traditionally the biggest landholders in the surrounding Gunnison Valley, initially resisted the trust's efforts. "The ranching community was very, very reluctant to be involved with the land trust," Bardeen says. "They thought it was just a tool to take their land away." But when they realized that donating conservation easements could help their families stay on their land—which was rapidly appreciating in value—by reducing estate taxes, many ranchers lent their support to the trust.

Since 1992, when the trust completed its first purchase of 11 acres along the Slate River, it has saved or helped to save more than 1,000 acres from development. In 1997, for example, the trust bought the Robinson parcel, 154 acres of highly visible open space between town and the ski mountain, for $1.2 million. More recently, the trust spent $1.9 million to preserve some 200 acres, including a trail linking the town to an outlying wilderness area.

"I really don't know of any opposition to the work we're doing with the land trust today," says Starr, a member of the land trust board. "Support for our work continues to grow."

For all its effort, the land trust can't buy out the whole valley, and today even Crested Butte is showing the signs of its success as a tourist mecca. New houses are springing up outside town, many of them huge structures built along ridgelines with no effort to hide them from view. "Unfortunately, many of the new houses are highly visible from town," Bardeen says. "But you can't stop growth. What the land trust is trying to do is consolidate development near areas that are already developed."

bike trail, nearby homeowners protested that it would encourage trespassing and threaten their privacy.[36] Similar concerns have been raised in many communities that have converted unused rail rights-of-way to recreational trails.

Land-use restrictions based on aesthetic considerations such as bans on ridge-line construction also generate intense debate. In a highly publicized case in Washington state, environmentalists went to court to force the builders of a new house overlooking the scenic Columbia River Gorge to move the house back because it violated regional building standards aimed at preserving the view for the public good. A state court upheld the regulation, but the owners said they would appeal.[37]

Some conservationists say state initiatives offer models for a shift in federal efforts to save open space. In Florida, for example, revenues raised by a land transfer tax imposed on real estate sales are used to purchase land for future use as parks and other public spaces. "We should be buying

Efforts to Save Evans Farm Came Too Late

Evans Farm has been a local landmark for decades in McLean, Va., an affluent Washington suburb. In fact, it was seen by many area residents as an oasis amid the surrounding suburban sprawl that included Tyson's Corner, a booming residential and business center. Open to the public for 40 years, the property included a Colonial-style restaurant, a replica of an old mill, a duck pond and meadows where horses and other farm animals grazed.

When Ralph Evans, owner of the 24-acre site, decided last year to sell the property to a local developer for a reported $20 million, area residents rallied to block the transaction. [1]

"This was a beautiful piece of land, with gently rolling hills, a lovely pond and many old trees," says Diane D'Arcy, an officer of the McLean Citizens Association. "It was a favorite site for rehearsal dinners, wedding receptions and anniversary parties. People from all over Northern Virginia came to the farm."

Dismayed at the prospect that a 144-unit housing development would replace the bucolic site, D'Arcy helped form the Coalition to Save Evans Farm. The group offered to raise enough money to buy the property outright. But Evans, whose family had owned the property since 1938, turned down the coalition's offer and accused its members of trying to force his hand.

"They never got serious about taking ownership of the property and instead used despicable tactics to try to block the deal, including lies and personal attacks," he says. "They acted as though my wife and I had awakened one morning and out of the blue decided to sell the farm out of greed and with no thought to the community. They knew nothing about the anguish involved in deciding to close down the business we'd run for 40 years."

In Evans' view, the coalition's goals had more to do with political interests than land conservation. "The motive was not to save open space but to promote the so-called smart-growth agenda, which in my view amounts to no growth," he says. "Planning development is the way to go, not just stopping development. In any case, the last I heard, we have a Constitution in this country that protects property rights."

Unable to convince Evans to change his mind, the coalition proposed creating a special tax district in McLean to raise money to buy the property and turn it into a public park, but the Fairfax County Board of Supervisors rejected the proposal. Coalition members then tried to block the rezoning permit required to develop Evans Farm, but county planners rejected their plea, saying the new development would comply with longstanding plans for the area. [2] When the developer, West Group, refused to scale back the project, coalition members organized a number of rush-hour protests in front of the developer's headquarters. In a final effort on July 28, residents pled their case before the Board of Supervisors. The board voted unanimously to let the development proceed.

West Group plans a mix of condominiums, townhouses and single-family houses on the site. The pond, the mill and some old trees will be saved, and the structures will be clustered to provide for open space within the development. But no recreational facilities are planned, and public access will be limited to sidewalks passing through the property.

From D'Arcy's point of view, the effort to save Evans Farm was crushed by a politically powerful developer. "Right from the get-go this was a political decision," she says. "All we were trying to do was buy the farm from Evans, and he called us communists and creeps who were trying to destroy his life. It was as nasty as can be."

For his part, Evans expresses relief but also bitterness about the experience. "I'll enjoy seeing the property developed the way West Group has planned," he says. "But I wouldn't want my worst enemy to go through what we did. For 40 years, we let the public use the property, maintained it at our expense and paid all the taxes. It just goes to show that no good deed goes unpunished."

Could Evans Farm have been preserved if a land trust had been involved in the negotiations? "It's impossible to say," says Russ Shay, public policy director of the Washington-based Land Trust Alliance, a national group representing groups and individuals involved in saving land. "But the most valuable lesson here is that people should recognize that if they're interested in preserving some aspect of open space in their community, they need to act early. Once a development agreement is made, it's very difficult to come in and undo that. They just got there too late."

[1] See Michael D. Shear and Peter Pae, "Evans Farm Loyalists Lose Out to Development," *The Washington Post,* July 28, 1999.

[2] See Michael D. Shear, "Evans Farm Vote Goes to Developer," *The Washington Post,* June 25, 1999.

more land, even if we mothball it for later designation, as Florida does," Pritchard says. "Making public lands our first priority would require our federal agencies to adopt a totally different orientation toward land conservation."

Property-rights advocates, however, say that Washington should look to recent state land-conservation efforts as an example of what to avoid in fashioning federal land policy.

"Smart-growth supporters suggest to people that their livability will go up if land-use controls are imposed around urban areas," Cushman says. "But forcing people into urban corridors will mean much higher population densities and much worse living conditions."

At the same time, Cushman warns, the value of rural property outside the urban growth boundary will plummet, depriving landowners of the true market value of their land. "There's always some good in any of these smart-growth proposals," he says, "but the overall effect will be negative for both rural and urban populations."

Land Trusts

The growth of private land trusts has paralleled the increased efforts of governments at all levels to save open space. Most of the nation's more than 1,200 trusts were organized in the past 15 years. Operating in all 50 states, trusts are especially active on the East and West coasts, where urban development has been especially intense.

Land trusts trace their roots to the "village improvement societies" that emerged in New England in the mid-1800s." [38] They spawned the Society for the Protection of New Hampshire Forests and other early trusts, many still in operation. Around the turn of the century, trusts were also created to save redwood forests in California and historical and archaeological sites in the Midwest.

Today's land trusts vary in size and scope, including local groups concerned with one or two critical properties, to regional trusts that coordinate open-space initiatives across state lines, to national groups like the Nature Conservancy, which has chapters in every state and has also collaborated with foreign groups to save more than 55 million acres overseas. Collectively, U.S. land trusts have helped protect 2.7 million acres from development, about half through outright land purchases and half through conservation easements.

Private land trusts are responsible for conserving open space in and around some of the most popular vistas in the United States, such as Big Sur on the California coast, the San Juan Islands in Washington's Puget Sound and lands abutting Acadia National Park in Maine and the Appalachian Trail.

More than half of all land trusts are true grass-roots organizations, with no paid officers; most of the remainder have only a director or a few staff members on the payroll. Nationwide, 70 percent of all funds used to purchase land come from contributions from members and individual donors. Some trusts receive money from foundations and corporations, while a number of states allocate funds to local land trusts for specific projects.

For example, Great Outdoors Colorado, a state agency, has funneled more than $35 million in state lottery funds to land trusts to protect some 60,000 acres of open space in the state. Land trusts also borrow money from banks, individuals and foundations to buy land; they repay the loans by selling the land to conservationists and through fund-raising drives.

Although most land trusts work independently and focus on local projects, a number of groups have begun pooling their efforts on regional initiatives. Land trusts are collaborating with local trail enthusiasts, for example, in the nonprofit East Coast Greenway Alliance, formed in 1991 to create a 2,500-mile bike and walking trail from Maine to Florida. [39]

The alliance's effort is only one of a number of ambitious conservation efforts by private land trusts. Last December the Wildlands Conservancy, a California trust, coordinated the biggest private land purchase in the state's history, a $52 million deal to buy more than 400,000 acres of privately owned land around Joshua Tree National Monument in Southern California.

In March, the New England Forestry Foundation announced the largest private forest conservation deal in U.S. history, a $30 million purchase of development rights to 750,000 acres in Maine's North Woods. [40] In September, the Conservation Fund brokered the purchase of 76,000 environmentally sensitive acres on the Delmarva Peninsula of Delaware, Maryland and Virginia, the largest conservation deal in the history of the Chesapeake Bay watershed.

Among the reasons land trusts have grown so rapidly, supporters say, is that they offer property owners a more efficient way to protect their land from development than trying to persuade local, state or federal agencies to buy the property itself or the rights to develop it.

"Of course, governments have far more resources to buy land, but land trusts are better to deal with than public agencies because they're generally more skilled in working out deals," says Shay of the Land Trust Alliance. "Because their mandate is to protect the public dollar, government agencies usually buy land for its assessed value, rarely for its full market value, whereas a charity will look for the best possible net return to the landowner."

Land trusts can save property owners time as well as money. "Once a government agency agrees to buy the land, it has to go to Congress to get the money to buy it," Shay says, "and that can take years."

Outlook

Action Stalled?

Congress appears unlikely to approve any major proposal affecting federal land conservation policy this year. Lawmakers already have gutted the president's Lands Legacy initiative, which was proposed as part of the $14.5 billion spending bill for the Interior Department. The conference agreement approved on Oct. 21 provided only $246 million of the $442 million Clinton had requested for the Land and Water Conservation Fund, a key item in the initiative. The approved funding level was down $82 million from last year's appropriation. Clinton is all but certain to veto the bill.

"This is a good bill," said John E. Peterson, R-Pa., during an Oct. 21 floor debate. "It is thoughtful; it has been a well-worked out compromise; it is the best we are going to get; and I think we should support it and the president should sign it." He added, "The agencies that are important to our environment have been thoughtfully funded."

But Rep. David R. Obey, D-Wis., ranking Democrat on the House Appropriations Committee, said the final version was not acceptable. "We feel that the conference report does not sufficiently take account of the opportunities available to us to save precious natural resources by meeting the president's request, or something close to it, for his Lands Legacy program," Obey said. "We have no choice but to stick by our convictions and oppose the bill at this point." [41]

The Interior appropriations bill also contains many anti-environmental riders, mostly benefiting resource industries. The riders include a delay in increasing the royalty payments that oil and gas drillers pay to extract resources from public property and a relaxation of limits on livestock grazing, logging and mining on federal land. Before the vote, Clinton warned he would veto the measure. "If the Interior bill lands on my desk looking like it does now," he said, "I will give it a good environmental response—I will send it straight back to the recycling bin." [42]

Conservation advocates are pinning their hopes on the bills still before Congress to fully and permanently fund the Land and Water Conservation Fund or expand tax incentives for landowners who voluntarily conserve their land. But they are frustrated at the apparent reluctance of lawmakers to approve any of the bills, especially those that offer tax incentives for saving open space.

"Under current law, a third of the country doesn't even qualify for the break in estate taxes, so expanding estate tax benefits would have a big impact on conservation," says Shay of the Land Trust Alliance. "It's a no-brainer." Although several bills aimed at promoting land conservation enjoy bipartisan support, no major bill has moved out of committee for full House or Senate consideration.

Critics say land-conservation proposals may backfire if governments become overzealous in imposing land-use regulations. "Open space and smart growth are slogans that politicians have coined for political benefit, but they have little foundation in reality," says Virginia developer Hazel, who adds that such efforts are ultimately doomed if they fail to respect suburban residents' needs and desires. "The people who need new roads and schools are already here, and they need to be provided for now, not by some future plan. The theory that you can just put them somewhere else is absurd, a fact that will resonate [among voters] sooner or later."

Despite congressional inaction thus far on land conservation bills, environmental advocates are confident that open-space efforts will continue to spread across the country, especially through support of local land trusts.

Fortunately, Shay says, no matter what governments do—or do not do—about land conservation, "a lot of people are pooling their resources to save open space. They are the ones who are looking to the future of their communities." ❖

Thought Questions

1. Should federal, state, and local governments place restrictions on where people can live? Why or why not?
2. If you were a developer designing a new community, what steps might you take to minimize negative aesthetic, geologic, and biologic impacts on the landscape?
3. On page 96 of this article (box entitled "Efforts to Save Evans Farm Came Too Late") farmer Ralph Evans, who has just sold his land for $20 million to a local developer, states "Planning development is the way to go, not just stopping development." Evans' statement reveals an underlying assumption that undeveloped land is wasted land. Is this a valid assumption? Why or why not?
4. What rights do other species—animal or plant—have in the face of development? What rights should they have?

FOR MORE INFORMATION

American Farmland Trust, 1200 18th St. N.W., Suite 800, Washington, D.C. 20036; (202) 331-7300; www.farmland.org. This private, nonprofit organization works to stop the loss of productive farmland and to promote farming practices that lead to a healthy environment.

Competitive Enterprise Institute, 1001 Connecticut Ave. N.W., Suite 1250, Washington, D.C. 20036; (202) 331-1010; www.cei.org. This public policy organization advocates free enterprise and limited government, and opposes federal land-conservation policies that include new purchases of private property.

Keep Private Lands in Private Hands Coalition, P.O. Box 400, Battle Ground, Wash. 98604; (360) 904-7472. The coalition of property-rights groups opposes federal government purchases of private land and was formed to fight land acquisition bills now before Congress.

Land Trust Alliance, 1319 F St. N.W., Suite 501, Washington, D.C. 20004-1106; (202) 638-4725; www.lta.org. The alliance represents more than 1,200 private land trusts around the country.

National Park Trust, 415 2nd St. N.E., Suite 210, Washington, D.C. 20002-4900; (202) 548-0500; www.parktrust.org. This nonprofit group works to preserve America's parklands. It buys inholdings, or privately owned land inside national parks, from willing buyers to slow development.

The Nature Conservancy, 4245 N. Fairfax Dr., Suite 100, Arlington, Va. 22203-1606; (703) 841-5300; www.tnc.org. A nationwide land trust with a million members, the conservancy operates the largest private system of nature sanctuaries in the world. Its sole focus is on land that provides habitat for imperiled species of plants and animals.

Notes

[1] Quoted by Steve Twomey, "Lots Not to Like," *The Washington Post*, July 5, 1999.

[2] For background, see Mary H. Cooper, "Setting Environmental Priorities," *The CQ Researcher*, May 21, 1999, pp. 425–448; Mary H. Cooper, "Urban Sprawl in the West," *The CQ Researcher*, Oct. 3, 1997, pp. 865–888; Tom Arrandale, "Public Land Policy," *The CQ Researcher*, June 17, 1994, pp. 529–552; and David Hosansky, "Traffic Congestion," *The CQ Researcher*, Aug. 27, 1999, pp. 729–752.

[3] See Sierra Club, "Solving Sprawl: The Sierra Club Rates the States," Oct. 4, 1999. See also Land Trust Alliance, "November 1998 Open Space Acquisition Ballot Measures," Feb. 12, 1999, which reports that voters approved 124 out of 148 initiatives for land acquisition only, raising a total of $5.3 billion.

[4] Land Trust Alliance, www.lta.org.

[5] For background, see Mary H. Cooper, "National Forests," *The CQ Researcher*, Oct. 16, 1998, pp. 905–928.

[6] From remarks on the Livability Agenda on Jan. 12, 1999.

[7] For background, see Kenneth Jost, "Property Rights," *The CQ Researcher*, June 16, 1995, pp. 513–536.

[8] Bureau of Land Management, "Public Land Statistics 1998," March 1999. In addition to the bureau, the Forest Service, the Fish and Wildlife Service and the National Park Service administer federal land.

[9] See The Wilderness Society Web site, www.wilderness.org.

[10] For background, see Richard L. Worsnop, "National Parks," *The CQ Researcher*, May 28, 1993, pp. 457–480.

[11] Babbitt spoke on Sept. 30, 1999, during a visit to the Chattahoochee National Recreation Area in Atlanta.

[12] Luntz Research Companies conducted the survey for The Nature Conservancy. See "American Views on Land & Water Conservation," summer 1999.

[13] Sierra Club survey conducted April 8–11, 1999.

[14] Sierra Club, "SPARE America's Wildlands," April 1999, p. 6.

[15] From a letter to Rep. Don Young, R-Alaska, and Sen. Frank Murkowski, R-Alaska, dated May 24, 1999.

[16] E-mail dated Aug. 21, 1999.

[17] See Michael Janofsky, "National Parks, Strained by Record Crowds, Face a Crisis," *The New York Times*, July 25, 1999.

[18] From a statement on Feb. 8, 1999.

[19] National Park Trust, "Saving the Legacy of the National System of Parks," Aug. 25, 1999.

[20] Hoeffel testified on Sept. 30, 1999, before a House Ways and Means

Committee hearing on the impact of tax law on land use, conservation and preservation.

[21] Johnson testified on Sept. 30, 1999, before the House Ways and Means Committee.

[22] "Vineyard Loophole," *The Wall Street Journal*, Aug. 24, 1999.

[23] Alexis de Tocqueville's classic, *Democracy in America*, was published in 1835–40.

[24] For background, see Mary H. Cooper, "Environmental Movement at 25," *The CQ Researcher*, March 31, 1995, pp. 283–306.

[25] See Kenneth M. Johnson, "The Rural Rebound," in *Reports on America*, Population Reference Bureau, August 1999.

[26] See William H. Frey, "New Sun Belt Metros and Suburbs Are Magnets for Retirees," *Population Today*, October 1999, pp. 1–3.

[27] From a statement of Aug. 25, 1999.

[28] Lloyd Billingsley, "Facts Versus Fantasy on Urban Sprawl," *Pacific Research Institute for Public Policy*, March 29, 1999.

[29] The act, which became law on Jan. 1, 1863, allowed anyone to file for a quarter-section (160 acres) of free land. Filers who built a house, dug a well, plowed 10 acres, fenced a specified amount and actually lived there gained title to the land after five years.

[30] In addition to buying private property, federal agencies often exchange low-priority public holdings for private property that is adjacent to parks, forests or other high-priority holdings.

[31] See Charles Babington, "Forest Protection Plan Is Unveiled," *The Washington Post*, Oct. 14, 1999.

[32] Sierra Club, "Solving Sprawl: The Sierra Club Rates the States," released Oct. 4, 1999.

[33] See David Kocieniewski, "Having Left Senate Race, Whitman Revels in the Job She Has," *The New York Times*, Sept. 12, 1999.

[34] See Mark Arax, "Putting the Brakes on Growth," *Los Angeles Times*, Oct. 6, 1999.

[35] See Scott Wilson and Susan DeFord, "Montgomery Bids for Open Space," *The Washington Post*, Oct. 14, 1999.

[36] See Patrick McMahon, "Residents Push to Derail Trails," *USA Today*, Oct. 7, 1999.

[37] See Stephanie Thomson, "Beas Appeal to High Court," *The Columbian* (Vancouver, Wash.), Oct. 8, 1999.

[38] Background information on land trusts from www.possibility.com/LandTrust.

[39] See Dieter Bradbury, "Greenway Fans Think Big: Maine to Florida," *Portland (Maine) Press Herald*, Sept. 26, 1999.

[40] See Pamela Ferdinand, "Private Deal to Preserve 750,000 Acres in Maine," *The Washington Post*, March 4, 1999.

[41] Charles Pope, "Senate Clears Interior Bill, Setting Stage for Post-Veto Talks on Policy Riders, Funding Levels," *CQ Weekly*, Oct. 23, 1999, p. 2527.

[42] From remarks during the Oct. 14 announcement of Clinton's new forest land policy.

Chronology

1800s

Federal land acquisitions set the stage for later government purchases of land for parks, national forests and wildlife preserves.

1803

Thomas Jefferson completes the biggest land deal in U.S. history when he pays France $23 million for the Louisiana Purchase, adding 530 million acres to U.S. territory.

March 1, 1872

Congress approves the creation of the nation's first national park, Yellowstone, a 2.2-million acre tract in northwestern Wyoming.

1940s-1960s

Postwar economic prosperity sets the stage for suburban development.

1949

The Federal Lands to Parks program is created to allow federal land to be transferred at no cost to state, regional and local governments for use in perpetuity as parks and conservation areas.

1951

The Nature Conservancy is founded to preserve threatened plant and animal habitat. It eventually becomes the nation's largest private land trust, with 1 million members.

Sept. 3, 1964

Congress establishes the Land and Water Conservation Fund and authorizes spending of up to $900 million a year on open space preservation. The fund is to be used for federal government land purchases and to provide matching grants to states for outdoor recreational projects.

1970s-1980s

State and local land planners begin taking steps to slow suburban "sprawl."

1973

Oregon's legislature requires all cities to contain most development within clearly defined urban-growth boundaries, providing a model for state and local land-conservation efforts.

1990s

The Clinton administration calls for new federal efforts to relieve traffic congestion and preserve open space.

September 1996

President Clinton designates 1.9 million acres of public land in southern Utah as the Grand Staircase-Escalante National Monument.

1997

Congress passes the Taxpayer Relief Act, which allows executors to reduce a landowner's estate-tax liability by donating a conservation easement on undeveloped property, which allows the land to be farmed or logged but never developed, even after changing hands.

Nov. 3, 1998

Voters approve more than 120 state and local ballot initiatives that provide about $5.3 billion in new funding for land conservation. Some measures create urban-growth boundaries to halt development beyond already-developed areas, while others raise taxes to buy land.

January 1999

President Clinton introduces his Lands Legacy initiative and asks Congress to give the Land and Water Conservation Fund its full annual funding of $900 million. Vice President Al Gore presents an administration proposal to provide $700 million in tax credits for states and local governments to build more "livable" communities.

August 1999

The Clinton administration agrees, pending congressional approval, to buy the Baca Ranch, a 95,000-acre tract in New Mexico's Jemez Mountains, for $101 million and turn it into a park.

September 1999

Federal officials complete the $13 million purchase of 12,000 acres of ranchland abutting Yellowstone National Park to be preserved as open space.

Oct. 14, 1999

President Clinton announces a plan to protect as much as 40 million acres of national forest land from development.

Oct. 21, 1999

Congress approves a fiscal 2000 Interior spending bill that slashes funding for the Lands Legacy initiative and other administration requests related to land conservation. The president threatens to veto the measure.

At Issue:

Should the federal government buy more land to help conserve open space?

PRESIDENT CLINTON

From his announcement of the Lands Legacy Initiative, Jan. 14, 1999.

Today I am proud to announce a Lands Legacy Initiative—$1 billion to meet the conservation challenges of a new century; . . . more than doubling our already considerable commitment to protect America's land. It represents the single largest annual investment in protecting our green and open spaces since [President] Theodore Roosevelt set our nation on the path of conservation nearly a century ago. And to keep on that path, we will be working with Congress to create a permanent funding stream for this purpose, beginning in 2001.

The first part of the plan builds directly on Theodore Roosevelt's conservation legacy by adding new crown jewels to our endowment of natural resources. Next year alone, we will dedicate $440 million, largely from the sale of oil from existing offshore oil leases, to acquiring and protecting precious lands and coastal waters. . . .

The second part of our plan, which works in tandem with the Livable Communities Initiative the vice president announced yesterday, represents a new vision of environmental stewardship for the new century. Today it's no longer enough to preserve our grandest natural wonders. As communities keep growing and expanding, it's become every bit as important to preserve the small but sacred green and open spaces closer to home—woods and meadows and seashores where children can still play; streams where sportsmen and women can fish; agricultural lands where family farmers can produce the fresh harvest we often take for granted.

In too many communities, farmland and open spaces are disappearing at a truly alarming rate. In fact, across this country, we lose about 7,000 acres every single day. And as the lands become more scarce, it becomes harder and harder for communities to then afford the price of protecting the ones that are left. That is why we have to act now.

So we will also dedicate nearly $600 million to helping communities across our country save the open spaces that greatly enhance our families' quality of life. With flexible grants, loans and easements, we will help communities to save parks from being paved over. We'll help to save farms from being turned into strip malls. We'll help them to acquire new lands for urban and suburban forests and recreation sites. We'll help them set aside new wetlands, coastal and wildlife preserves. There will be no green mandates and no red tape. Instead, the idea is to give communities all over our country the tools they need to make the most of their own possibilities.

MALCOLM WALLOP—Chairman, Frontiers of Freedom; former Republican senator from Wyoming

From testimony before the Senate Energy and Natural Resources Committee, May 4, 1999.

Government doesn't need to own any more land. It already owns far too much land and far more than it can take care of properly. The four federal land agencies, according to the [Bureau of Land Management], control about 676 million acres, or [more than] 29 percent of the country. Other federal agencies, such as the Department of Defense, own more millions of acres. State and local governments also own a lot of land, although no one knows exactly how much. . . .

I don't think a plausible case can be made that government needs to own even more land for the purpose of environmental protection. A much stronger case can be made that private owners provide, on average, a much higher level of environmental stewardship than does public ownership and, therefore, that the environment would be healthier if we had less public land rather than more. . . .

Before the Congress embarks on a land-buying spree, it would seem to me prudent to consider how all this new property is going to be maintained. There are only two alternatives. Either taxpayers are going to have to pay more to maintain these new acquisitions, or the current appropriation must be stretched ever thinner to maintain more and more land. The budget for Yellowstone National Park and many of our other great national treasures is not adequate now. Adding more land within the current budget constraints means even less money for Yellowstone. . . .

Removing hundreds of millions of dollars of private land from productive uses every year would significantly reduce economic activity in many areas and consequently reduce the tax base. After that is accomplished, these state and local governments would then be burdened with the cost of maintaining their public lands. . . .

The American system of constitutionally limited government was instituted in order to secure the blessings of life, liberty and property to all citizens. The revolutionaries of 1776 and the delegates to the Constitutional Convention of 1787 were very familiar with the old political maxim that "power follows property." The more property government owns or controls, the less power the people retain. As the balance of power shifts towards government, the more difficult it becomes for the people to maintain the blessings of life, liberty, and property. For this reason, the founders would have opposed our vast federal estate just as surely as they would have opposed our confiscatory levels of taxation.

Bibliography

Selected Sources Used

Articles

"Urban Sprawl: Not Quite the Monster They Call It," *The Economist*, Aug. 21, 1999, pp. 24–25.
Federal policies, such as generous spending for highway construction and tax breaks to encourage home ownership, have encouraged suburban development. Sprawl also is the result of an express desire on the part of middle-class Americans to leave congested cities for better living conditions.

Allen, Jodie T., "Sprawl, from Here to Eternity: It's Maddening as Hell. But What Can Washington Really Do about It?" *U.S. News & World Report*, Sept. 6, 1999, pp. 22–23.
Vice President Al Gore's Livable Communities initiative is criticized as "green pork" because it would preserve coastal regions in Alaska and other areas with powerful representatives in Congress.

Davis, Tony, "Tucson Paves Its Way across a Fragile Landscape," *High Country News*, Jan. 18, 1999, p. 1.
Like other Arizona communities, Tucson is growing fast to accommodate retirees and other new residents. Although growth continues, efforts are now under way to contain sprawl through land purchases and stricter zoning.

Gurwitt, Rob, "The State vs. Sprawl," *Governing*, January 1999, pp. 18–23.
Maryland Gov. Parris N. Glendening's "smart-growth" policy limiting most development to existing communities is being closely watched by other state officials to determine whether it can serve as a model for their own efforts to alleviate traffic congestion.

McChesney, Jim, "Portland: Urban Eden or Sprawling Hell?" *Oregon Quarterly*, summer 1999, pp. 18–23.
State urban-growth mandates introduced in 1973 have forced Oregon's biggest city to contain growth close to existing development. While many residents and planners praise the result of these policies, critics say they have merely raised real estate values, pricing many lower-income people out of the city.

Reports and Studies

"Building Livable Communities: A Report from the Clinton-Gore Administration," June 1999.
The report describes how open-space loss results from suburban development and how administration proposals would save undeveloped land.

General Accounting Office, "Community Development: Extent of Federal Influence on 'Urban Sprawl' Is Unclear," April 1999.
The independent investigative agency for Congress concludes that efforts to regulate land use to combat sprawl traditionally have come under the jurisdiction of state and local governments and that the eventual impact of closer federal involvement in this area is uncertain.

Johnson, Kenneth M., "The Rural Rebound," Population Reference Bureau, August 1999.
For most of the nation's history, people have migrated from the countryside to cities in search of employment. Computer technology and economic prosperity have recently fueled a reversal of the trend, posing new policy issues for land-use planners.

Riggs, David, and Daniel Simmons, "Anti-Sprawl Policy: Congested Thinking and Dense Logic," policy brief, Competitive Enterprise Institute, Aug. 9, 1999.
Growth-management policies are ineffective and little more than nostalgic attempts to recreate outmoded communities, according to this brief for CEI, a public policy group that advocates free markets and a limited role for government.

Sierra Club, "1999 Sierra Club Sprawl Report," Oct. 4, 1999.
The San Francisco-based environmental group rates the states in terms of their efforts to protect open space and emphasis on land-use planning. Maryland and Oregon receive the group's highest ratings.

Staley, Samuel R., "The Sprawling of America: In Defense of the Dynamic City," Reason Public Policy Institute, undated.
Less than 5 percent of the nation's land is developed, Staley writes. He argues that concern about suburban sprawl is overblown and that, in any case, development is a matter for local officials, not the federal government, to regulate.

United States Geological Service, "Perspectives on the Land Use History of North America: A Context for Understanding Our Changing Environment," 1998.
The Interior Department agency presents an exhaustive study of trends in land development across the country, including urban growth and agricultural production.

The Next Step

Additional information from UMI's Newspaper and Periodical Abstracts™ database

Preserving Open Spaces

"Champion Completes Sale of Timberland to Conservation Fund," ***The Wall Street Journal*, July 2, 1999, p. B2.**
The forest products company Champion International Corp. says it has completed the sale of 143,000 acres of New York timberland to the Conservation Fund, a nonprofit group, for $46 million.

Clifford, Frank, "San Diego OKs Land Conservation Plan; One-quarter of a 900-square-mile Area Would be Protected, While Guaranteeing Developers 'No Surprises' Under Endangered Species Act," ***Los Angeles Times*, Oct. 23, 1997, p. A3.**
The San Diego County Board of Supervisors approved the nation's most ambitious urban land conservation strategy, designed to protect 85 imperiled species of plants and animals across 900 square miles of Southern California. The San Diego plan softens the impact of the Endangered Species Act on new development while attempting to protect open space and ensure the survival of wildlife facing extinction along one of the fastest-growing real-estate corridors in the country.

Ferdinand, Pamela, "Three States Reach Land Conservation Deal; 300,000 Northeast Acres Set for Logging, Public Use," ***The Washington Post*, Dec. 10, 1998, p. A3.**
New York, Vermont and New Hampshire joined forces with environmentalists and private investors to place 300,000 acres of the Northern Forest into a $76.2 million public-private conservation project hailed as the largest in U.S. history. The deal will keep the vast majority of the woodlands intact as a traditional working forest available for regulated logging but also open for public recreation.

Franklin, James L., "Group Aims to Preserve 455 Parcels as Open Space," ***The Boston Globe*, July 18, 1999, p. WKW1.**
A volunteer group in Massachusetts is trying to preserve open spaces through land acquisition, conservation easements and cooperation with towns and institutions. The Trustees of Reservations owns or manages 82 properties, with 21,300 acres. It also protects an additional 11,700 acres through conservation restrictions on more than 175 parcels of privately owned land. Overall, the Trustees hope their work with private landowners, other nonprofits and government can help the state protect 200,000 acres by 2010, a 100 percent increase in the amount of protected open space in the state.

Harris, Elaine, "Report Shows Voter Support for Land Conservation," ***Nation's Cities Weekly*, Feb. 22, 1999, p. 10.**
Livability and smart-growth initiatives are not only topping the national policy agenda, but state and local agendas as well. A recent report found 240 state and local initiatives on conservation and smart growth on November ballots.

Miniclier, Kit, "Easement Keeps Land in Family; Rancher Ritter Saves 'Million-dollar View,'" ***The Denver Post*, June 4, 1999, p. B1.**
In order to save their family's century-old ranch and their view of the Spanish Peaks, Gail Ritter, 69, and her 89-year-old cousin, Claudia Capps, have placed 1,750 acres of the family ranch in a conservation easement, legally barring forever its sale for real estate or other commercial development by signing over development rights to the Colorado Cattlemen's Agricultural Land Trust.

Morley, Jefferson, "Owens Outlines Plan to Save County's Farmland," ***The Washington Post*, April 29, 1999, p. MDA1.**
Anne Arundel County (Md.) Executive Janet S. Owens announced a plan to triple county spending on farmland preservation, offering farmers incentives not to sell their land to developers. The program, called the Installment Payment Alternative, would enhance the county's current program to put rural properties under permanent conservation easements, ensuring that land can never be used for commercial or residential development.

Land Trusts

Bayles, Fred, "Land Trusts Embrace Grass Roots; More Groups Take up Cause on Local Level," ***USA Today*, Oct. 1, 1998, p. A3.**
The Land Trust Alliance, a Washington, D.C.-based coalition of private land trusts, released a study showing a dramatic rise in land preservation at local and regional levels. Holdings of smaller groups remain a fraction of the land saved by national conservation groups. But Jean Hocker, president of the Land Trust Alliance, says the growing power of smaller land trusts speaks to increasing local activism focused on saving nearby open lands from development.

Brooke, James, "Land Trusts Multiplying, Study Shows," ***The New York Times*, Oct. 1, 1998, p. A20.**
Land trusts, once restricted to genteel corners of New England, are

now growing the fastest in the Rockies and the Southwest. In the last decade, as ranchers and urbanites grafted the land-trust movement onto the rural West, the number of local land trusts jumped by 160 percent in the five-state Rockies division, 147 percent in the four-state Southwest division and 108 percent in the six-state Western division.

Cornelius, Coleman, "Land Trusts Seek to Save Open Space; Groups Resisting Tide of Development," *The Denver Post*, May 2, 1998, p. B6.
The state's land trusts have been key players in identifying valued land and using a variety of tools to preserve it. Members of statewide land trusts, city and county planners and others concerned about open space met in Fort Collins, Colo., to share information they hoped would promote successful preservation efforts. In 1993, there were 140,000 acres of land in Colorado that had been protected by 26 land trusts through open-space purchases or conservation easements, according to a coalition survey. By 1996, that number had doubled, to 280,000 acres of protected land.

Wood, Daniel B., "Nature Preserves Proliferating in U.S.; Thriving Grass-roots Land Trusts Have More than Doubled the Amount of Land Set Aside for Nature During the Past Decade," *The Christian Science Monitor*, Oct. 1, 1998, p. 1.
The idea behind land trusts dates back decades, but the past 10 years have seen remarkable growth in the amount of land that grass-roots groups have socked away. Since 1988, trusts have more than doubled the amount of protected land, from 2 million to 4.7 million acres—an area larger than Connecticut and Rhode Island. Citizens with no prior knowledge of how to preserve land by purchase, donation and tax incentive are joining neighbors to do just that. The number of the groups themselves has also soared.

The Nature Conservancy

"Group Buys Chunk of Maine Forest; Nature Conservancy Will Open About 185,000 Acres of Privately Owned Land to Public After $35.1-million Deal, Officials Say," *Los Angeles Times*, Dec. 16, 1998, p. A14.
About 185,000 acres of privately owned Maine wilderness will be opened for public recreation in a $35.1-million conservation deal. The purchase by the Nature Conservancy from International Paper Co., the largest of its kind in Maine's history, comes less than a week after New York, Vermont and New Hampshire joined with private investors and environmentalists in a $76-million deal to preserve 300,000 acres of similar wilderness. The 286 square miles of unbroken Maine forest includes a 40-mile stretch of the Upper St. John River in the state's far northwestern corner, an area heavily populated by moose and bear. It is also home to the second-highest concentration of rare plants in Maine.

"Nature Conservancy of Texas Shows How It's Done," *Houston Chronicle*, Oct. 31, 1998, p. A36.
The state chapter of this private, nonprofit group has acquired more than 365,000 acres of unique Texas habitat, from Galveston marshes to the Davis Mountains in West Texas. The Conservancy works with public agencies to protect wildlife along Armand Bayou and on the Katy Prairie in the Houston area and in numerous other projects throughout the state.

Smith, Leef, "Garden for Learning: Nature Conservancy Creates Suburban Oasis of Unusual Plants," *The Washington Post*, June 10, 1999, p. VAW1.
The Nature Conservancy chose almost 100 varieties of nearly extinct plants—each native to the United States—to be planted in the back yard of its new international headquarters.

Using Public Land

Decker, Peter R., "Striking a Balance in Use of West's Public Land," *The Denver Post*, Oct. 4, 1998, p. L1.
The author addresses the growth of the tourism and recreation industries and their impact on the use of public land. He says those who want to preserve public lands for recreational purposes should not have more say in land use than those who, by choice or necessity, must earn a living on these lands.

Knickerbocker, Brad, "New Twists in Debate Over Public Land; 'Sagebrush Rebellion' is Stoked by a Senator's Proposal to Sell More Federal Land Each Year," *The Christian Science Monitor*, April 23, 1998, p. 4.
Conservative Western lawmakers bristle at the idea of greater environmental protections for public lands—lands that are traditionally devoted to the extracting of resources by private interests. When Forest Service chief Michael Dombeck announced a moratorium on road-building in national forests, Sens. Frank Murkowski, R-Alaska, and Larry Craig, R-Idaho, as well as Reps. Don Young, R-Ala. and Helen Chenoweth, R-Idaho, threatened to severely cut the agency's budget.

Murphy, Kim, "Alaska's Delicate Arctic Awaits New Push for Crude; Exploratory Oil Drilling Now Seems Fated for the Biggest Patch of Untouched Public Land in the U.S.," *Los Angeles Times*, July 10, 1997, p. A1.
A year after conservationists beat back attempts to begin drilling in the nearby Arctic National Wildlife Refuge, a showdown is looming in Alaska's 23.4 million-acre National Petroleum Reserve—a contest that the oil companies appear likely to win. At a time when the trans-Alaska pipeline and the continent's largest oil field at Prudhoe Bay was supposed to have been tapped out, a stunning array of North Slope finds means a production field that has been in decline since the late 1980s will open the taps even wider over the next several years.

Geologist M. King Hubbert once said that he found it hard to decide which fact is more remarkable, "that it took 600 million years for the Earth to make its oil, or that it took 300 years to use it up" (see Press & Siever, pp. 510–511, for a discussion of how oil is formed and where it is found; Merritts, De Wet, & Menking, pp. 331–334, 335–336). That the world is running out of oil is indisputable. With an annual consumption of 6.8 billion barrels as of 1998, the United States leads the world, using 25% of the total annual supply (see Press & Siever, p. 512, pp. 116–117 of this article; Merritts, De Wet, & Menking, p. 334). Sixty percent of that consumption is used in transportation, and the recent sport utility (SUV) and light truck craze has led to acceleration in the rate of oil use. So too has the trend toward increasing suburban sprawl, which makes people dependent on their cars to reach their workplaces, schools, and shopping centers. As the U.S. demand for oil grows, so too does our dependence on foreign oil imports, a fact that has serious implications for our economic health. As Mary Cooper notes in the following article, overdependence on foreign oil imports led to a severe recession in the 1970s as members of the Organization of Petroleum Exporting Countries (OPEC) imposed an embargo against the United States for its support of Israel in the Yom Kippur war. After several painful years in which the price of oil skyrocketed by more than 300%, the U.S. government began to invest more in local exploration for oil and in development of alternative energy resources. Repeated crises throughout the 1970s, 1980s, and early 1990s led to fluctuations in the price of oil, but the Asian financial crisis of the late 1990s caused a decrease in industrial consumption in that part of the world, leading to a glut of oil on the world market. Oil prices fell to their lowest inflation-adjusted values in 30 years.

The low price of oil has allowed gas guzzling SUVs and light trucks to become the most popular vehicles in America today, with more than 50% of new car sales falling into these categories. In addition, tax incentives developed during the Carter administration to support the use of renewable energy resources like wind power and solar energy were repealed under the Reagan administration. Since this article was written, oil prices have again shot up, as OPEC nations have decided to limit production to recoup their losses from the oil glut of the late 1990s. The United States again finds itself in a difficult position, and has high hopes that the oil-rich Caspian Sea region may help us to reduce our dependence on imports from OPEC nations. Though this region appears to be resource rich, political turmoil and ethnic strife make exploitation of these resources uncertain. Political problems aside, there are other reasons why the United States should take steps to decrease its dependence on foreign oil imports by widely developing renewable resources. Fossil fuel combustion creates the greenhouse gas carbon dioxide, which appears to be causing global warming and changes in the global climate system (see Press & Siever, pp. 562–566; Merritts, De Wet, & Menking, pp. 401–404). As Earth warms, the sea level is rising, threatening coastal communities. In addition, computer models predict that the hydrologic cycle will change, increasing the possibility of drought in some locations and resulting in greater frequency of hurricanes and tropical storms. Extraction and consumption of oil as currently practiced also leads to severe ecologic damage. Oil spills such as the Exxon Valdez disaster (see Press & Siever, p. 511; Merritts, De Wet, & Menking, pp. 337–338) are responsible for killing sensitive coastal communities. Hundreds of thousands of shore birds have been killed by oil pollution as have countless mollusks, crustaceans, fish, and other aquatic species. Development of renewable energy resources promises not only to enhance our energy self-sufficiency, but also to help us to tread more lightly upon the Earth (see Merritts, De Wet, & Menking, pp. 350–359, for a discussion about alternative energy sources and conservation).

Oil Production in the 21st Century

When will the world run out of oil?

Twenty-five years ago, the Organization of Petroleum Exporting Countries struck at the heart of the American economy with an embargo on oil exports to the United States. The resulting rise in energy prices sparked a round of inflation and stagnant economic growth that lasted more than a decade. A quarter-century later, gasoline prices are at an all-time low, new oil deposits have been discovered in the Caspian Sea region and OPEC appears to have lost its grip on global energy prices and production. But the good times for consumers will not last forever. In a matter of time the world's oil will run out, and it's far from certain there will be sufficient alternative energy sources.

North Sea oil-drilling platfrm (Public Broadcasting Service)

THE ISSUES

109 • Has OPEC lost its control over global oil prices?
• Does U.S. foreign policy enhance Americans' access to foreign oil?
• Will the United States be ready with alternative energy sources when the oil runs out?

BACKGROUND

117 **OPEC's Power Play**
The Organization of Petroleum Exporting Countries was founded in 1960 to control oil prices.

CURRENT SITUATION

119 **OPEC's Dilemma**
The oil cartel's ability to control the global oil market in the future has been questioned.

119 **Cheap Oil**
Most oil-producing nations are suffering from lost revenues.

122 **Caspian Treasure**
Oil companies are hotly competing for the huge reserves under and around the Caspian Sea.

123 **Pipelines and Politics**
Political turmoil in the Caspian region has slowed development of new oil pipelines.

OUTLOOK

124 **Will Prices Rise?**
Oil prices could remain low after the current glut vanishes because the world is weaning itself from petroleum products.

SIDEBARS & GRAPHICS

110 **Caspian Oil Depends on Pipelines**
Several new routes have been proposed to move oil to international markets.

111 **Crude Oil Prices Have Been Dropping**
OPEC's control has been challenged.

113 **U.S. Dependence on Imported Oil Growing**
Falling domestic production reflects decreasing U.S. reserves.

114 **Getting Oil Out of the Caspian**
Political turmoil in the region could slow progress.

116 **U.S. Taps Varied Oil Sources**
Four of the five top U.S. imports are not Arab nations.

120 **As Oil Runs Out, Technology Buys Time**
Advances in oil exploration and extraction could raise production by 20 percent.

127 **Chronology**
Key events since 1960.

128 **At Issue**
Should the U.S. ease its sanctions against Iran to improve international access to Caspian Sea oil?

FOR FURTHER RESEARCH

129 **Bibliography**
Selected sources used.

130 **The Next Step**
Additional sources from current periodicals.

Note: For more information on this topic, please see the following pages in Press and Siever's *Understanding Earth*, Third Edition: pp. 510–511, discussion of how oil is formed and where it is found; pp. 512, 562–566; and in Merritts, De Wet, and Menking's *Environmental Geology:* pp. 331–334, 335–336, 337–338, 350–359, 401–404.

Oil Production in the 21st Century

By Mary H. Cooper

The Issues

Vacationing Americans have an extra reason to celebrate this summer: At less than $1 a gallon in some areas, gasoline prices are at their lowest level in 30 years.

Several factors produced this happy turn of events. An unusually warm winter in the Northern Hemisphere, caused in part by El Niño's disruption of normal weather patterns, dampened demand for heating oil. At the same time, a severe economic crisis in East Asia forced Japan and other countries in the region to curtail industrial production, thus reducing energy consumption.

Faced with falling demand for oil, the Organization of Petroleum Exporting Countries (OPEC) and other oil producers tried to prop up global oil prices by limiting production. But with oil prices falling, many producers ignored calls to slow production in a desperate attempt to protect oil revenues.*

Just a quarter-century ago, however, OPEC had the United States and the rest of the industrialized world by the throat, and Saudi Arabian oil minister Sheik Zaki Yamani was as familiar to many Americans as their representatives in Congress. Saudi Arabia and the four other major Middle East members dominated OPEC, controlling more than a third of world oil production.

In October 1973, the cartel imposed a total embargo on its exports to the United States and the Netherlands for their support of Israel in the Yom Kippur War. Oil prices skyrocketed. The ensuing inflation and stagnant industrial output spilled over from the U.S. and Dutch economies to infect the entire industrial world. Stagflation became even more deeply entrenched after 1978–79, when the Iranian revolution sparked a second energy crisis and rise in oil prices.

The industrial world responded to OPEC's grip on world oil supplies by searching for alternative sources. By the 1990s, non-OPEC producers such as Britain, Mexico and Norway had enabled importers to reduce their oil dependence on the volatile Persian Gulf. They also launched campaigns to reduce their consumption of oil by raising energy taxes and encouraging energy conservation through improved efficiency of automobiles and appliances. And they sought to develop alternatives to oil such as solar, wind and geothermal energy.

These efforts paid off in the next two decades. Largely by diversifying their sources of foreign oil, the United States and other major consumers have significantly reduced their vulnerability to oil price manipulations by OPEC. Technological advances in exploration and drilling equipment have enabled producers to discover new oil deposits, further relaxing pressure on oil prices posed by growing global demand for oil.

Worries about future energy crises diminished still further when oil reserves under and around the Caspian Sea were opened to outside development after the Soviet Union's collapse in 1991. The former Soviet republics of Azerbaijan, Kazakhstan and Turkmenistan that border the Caspian—and, to a lesser extent, their

*OPEC members are Algeria, Indonesia, Iran, Iraq, Kuwait, Libya, Nigeria, Qatar, Saudi Arabia, the United Arab Emirates and Venezuela.

Getting Oil Out of the Caspian

Several new pipelines have been proposed to move oil from the landlocked Caspian republics to maritime ports for export. The U.S. favors lines running from Baku, Azerbaijan, to Ceyhan, Turkey, or from Baku to Georgia. Russia prefers a northern pipeline connecting to its own system or to its Black Sea port of Novorossisk. The pipeline sought by the Caspian republics, runnning south to the Persian Gulf, is opposed by the U.S. because it crosses Iran. The route planned by China across Kazakhstan is not controversial.

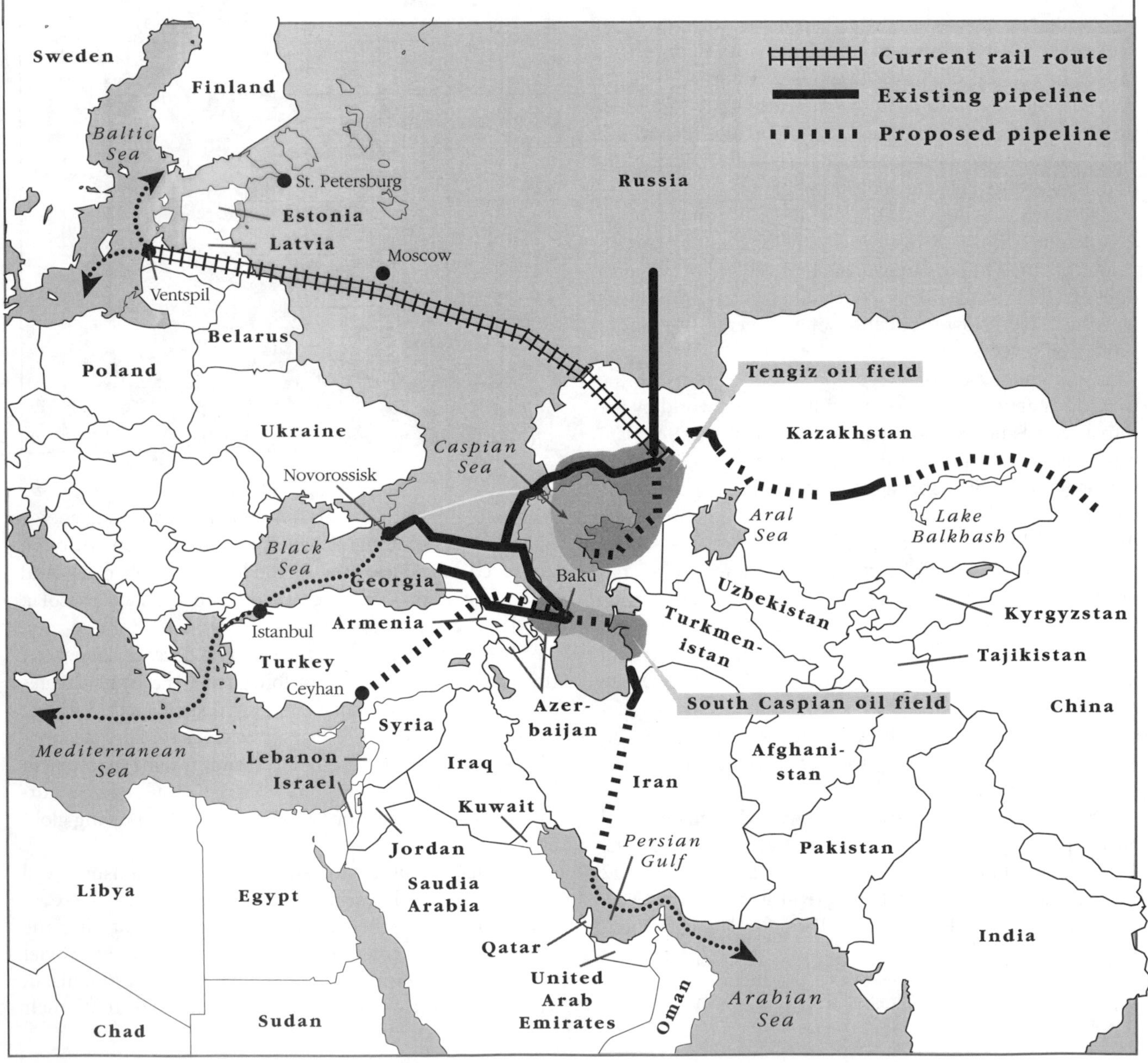

Sources: Energy Information Agency, Fortune Magazine, Parade Magazine

Central Asia neighbors Tajikistan and Uzbekistan—stand to reap enough earnings from oil and natural gas exports alone to launch them into the modern industrial era in the space of a few years.

"The Caspian is potentially one of the world's most important, new, energy-producing regions," said former Energy Secretary Federico Peña earlier this year. "Although the Caspian may never rival the Persian Gulf, Caspian production can have important implications for world energy supplies by increasing world supply and diversifying sources of supply among producing regions of the world." [1]

Crude Oil Prices Have Been Dropping

World crude oil prices have dropped in recent years as members of the Organization of Petroleum Exporting Countries have resisted OPEC efforts to limit their oil production. Discoveries of new oil sources in non-OPEC nations also have kept prices down.

Source: Energy Information Administration, Annual Energy Review, *1997*

The oil glut that began early this year provided the icing on the cake for oil consumers. But Americans would be foolish to gloat over their current bounty. Even counting Alaska's prolific North Slope oil fields, domestic reserves are falling. The United States now imports more than half of its oil, placing the country at added risk from future disruptions of foreign supplies. [2]

For all its promise, the Caspian Sea region is far from coming on line as a major source of non-OPEC oil. The region's remoteness and legendary political turmoil threaten to postpone or even scuttle the flow of oil before production is fully under way.

Meanwhile, low gasoline prices have lulled American motorists into trading in their energy-efficient sub-compact cars for gas guzzling sport utility vehicles, which now account for almost half of American new car sales. Programs to develop alternatives to oil are losing support in Congress, which has cut the Energy Department's research budget.

America's growing oil consumption also flies in the face of concern that using oil and other fossil fuels is causing a gradual but potentially catastrophic warming of Earth's atmosphere. The United States joined 167 other countries last December in agreeing to reduce fossil fuel consumption. But in the current climate of energy abundance, support for the Kyoto Protocol is flagging, and the Senate appears highly unlikely to ratify the measure anytime soon.

Although consumers may rejoice in OPEC's recent inability to curtail output and raise prices, it may be too soon to write the organization's obituary. OPEC has begun seeking agreement from non-member producers to go along with its efforts to buoy sagging prices. In March, Mexico agreed to join OPEC in cutting production, prompting some oil experts to predict that other countries soon would join in. [3]

Even absent a strengthening of OPEC's ability to manipulate oil prices over the short term, some experts contend that the recent break in oil prices will prove to be temporary. Colin J. Campbell, an oil-industry consultant in Geneva, Switzerland, and author of *The Coming Oil Crisis*, predicts that prices will rise when global oil production reaches its peak, "within the first few years of the next century." After that, he predicts, demand for oil will outpace its supply. "This will be a fundamental turning point, because until now we've always had growing oil production," he says.

As global oil deposits are depleted, OPEC's Middle East producers could be left in control of an increasing portion of world reserves. "We can expect another major price shock around 2000," Campbell says, "when the Middle East's share of world reserves will be much greater than it is now."

As Americans fire up their gas guzzlers for summer outings, these are some of the questions oil experts are asking:

Has OPEC lost its control over global oil prices?

Through new discoveries and an expanded membership, OPEC's oil reserves have grown over the years. The organization has reached beyond its stronghold in the Persian Gulf—still the main source of the world's oil—to include far-flung producers such as Indonesia and Nigeria. And OPEC's undisputed leader, Saudi Arabia, retains its clout as the world's largest oil producer.

But OPEC's grip over world oil production has slipped in the past 20 years, largely as a result of its own actions. When it set strict quotas in the 1970s that quadrupled oil prices, the cartel sparked a frantic search for alternative sources of oil by the industrial world, which depends on petroleum products for its economic survival.

With the exception of the United States, most industrial countries imported the bulk of their energy supplies. Using sophisticated technology, Britain and Norway soon located and began working oil deposits under the North Sea. *(See story, p. 120.)* The United States and other countries shifted much of their oil demand to these and other non-OPEC producers, which now account for 45 percent of the oil export market. Norway now is the world's second-largest exporter, after Saudi Arabia.

Just how much has OPEC lost its ability to set production quotas among its membership as a way to buoy prices? That became apparent last fall, when a global oil glut began depressing prices. As usual, Saudi Arabia took the lead, calling in March for a cut in oil production of 1.2 million barrels a day. But with prices falling—and with them precious oil revenues—many members ignored their quotas and pumped as much oil as the market would bear.

Desperate to reduce output, OPEC has appealed to other producers, which are also feeling the pinch of falling oil revenues. Russia, which has the largest oil reserves of any non-OPEC country and badly needs oil revenues to stave off its deepening financial crisis, attended OPEC's June 24 meeting in Geneva and may soon join the organization. [4] On June 4, the oil ministers of Saudi Arabia and Venezuela reached an agreement with Mexico to cut production by 450,000 barrels a day, beginning July 1. This agreement has had some effect on prices, leading some observers to predict that OPEC will turn more often to informal agreements of this kind in an effort to regain leverage over the market and stabilize prices around the historical norm of about $20 a barrel.

"OPEC is a major factor in determining world oil prices," says Edward H. Murphy, director of finance, accounting and statistics for the American Petroleum Institute (API), which represents U.S. oil companies. "They have succeeded in reducing production by about 2.1 million barrels a day, or 6 percent, since February. That is a significant factor that has prevented oil prices from falling even further. They're not back up to $20 a barrel, but they'd be a lot lower today in the absence of OPEC's production cuts."

Does U.S. foreign policy enhance Americans' access to foreign oil?

A fundamental, though infrequently recognized, goal of U.S. foreign policy has been to ensure the access of American businesses and consumers to foreign oil supplies, especially since 1971, when domestic oil production peaked and began its gradual decline. By 1996, the United States—once a leading exporter of crude—was for the first time forced to import half of its oil. Today, imports account for 52 percent of U.S. consumption. *(See graph, p. 113.)* As demand for oil continues to rise, the United States will likely depend on imports for an ever-growing portion of its oil supply.

Like other industrialized countries, the United States has diversified its sources of foreign oil away from the Middle East since the energy crises of the 1970s. Today, Venezuela is the leading source of U.S. oil imports, followed by Canada. Although U.S. dependence on Persian Gulf oil has fallen from 28 percent of oil imports in 1991 to just over 19 percent today, the region will remain a vital oil supplier for years to come. But it is also one of the most politically unstable regions, the focus of 50 years of hostilities between Israel and its Arab neighbors and, for the past two decades, of militant Islamic fundamentalism.

Access to oil has figured prominently in the United States' activities in the Middle East throughout this period, most recently in the 1991 Persian Gulf War, when the United States led a United Nations military coalition that forced Iraq to withdraw from neighboring Kuwait. [5]

Of course, access to oil is not the sole U.S. strategic interest in the region. Even during the gulf war, the Bush administration stressed the need to repel Iraq's invasion of Kuwait to maintain the international rule of law. Because Iraq's leader, Saddam Hussein, was suspected of producing nuclear and biological weapons, the United States also sought and obtained Security Council support for a U.N. embargo against Iraq pending the completion of U.N. inspections of Iraqi arsenals. But the Clinton administration's "dual containment" policy toward Iraq and Iran—which together control 20 percent of the world's proven reserves—has implications for U.S. access to the region's oil exports. [6]

Iraq, with 112 billion barrels of proven oil reserves, is second only to Saudi Arabia, with 262 billion barrels. The embargo has exacted a heavy toll on Iraq, depriving it of oil revenues and leaving many Iraqis without adequate food and medical supplies. In December 1996, the embargo was eased to allow Iraq to export just enough oil to pay for food and other essential supplies. This "oil-for-food" provision is to remain in place until the Iraqi government allows U.N. arms inspectors full access to weapons sites.

But some oil-importing countries, notably France, Russia and China, oppose continuing the embargo against Iraq indefinitely. Critics of the embargo welcomed U.N. Secretary General Kofi Annan's role in prodding Saddam Hussein to allow U.N. inspectors greater access to the country last February. Annan signaled a softening of international attitudes toward Iraq by calling the Iraqi leader "a man you can do business with." For his part, Saddam has threatened to take unspecified action to break the embargo if it is not lifted this year. [7]

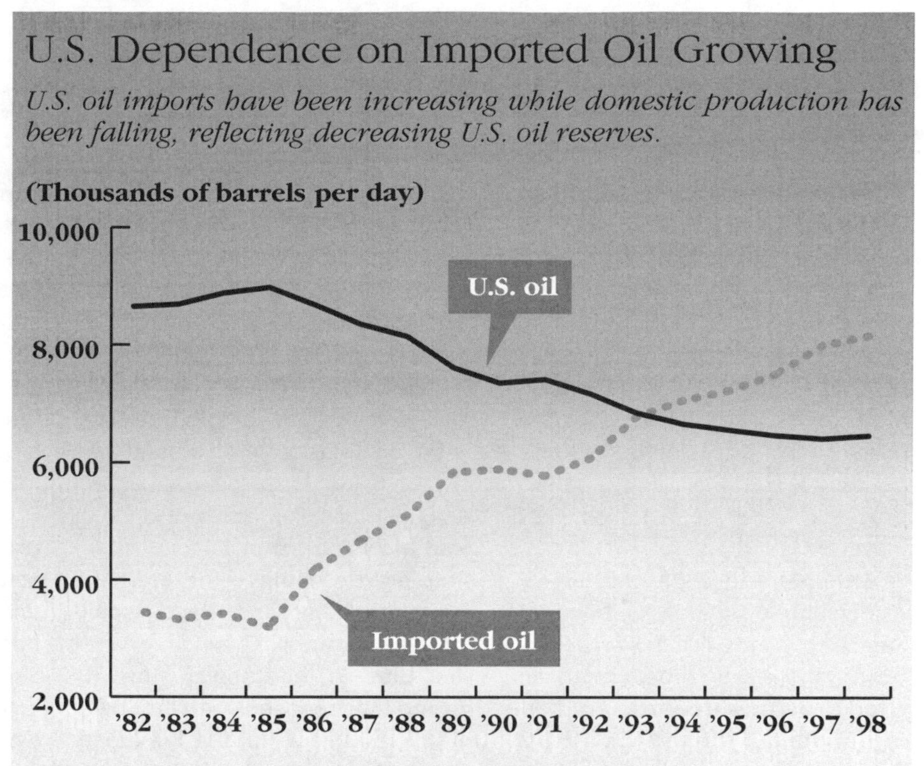

Indeed, Iraq is already expanding oil exports in defiance of the U.N. embargo. "The fact that Iraq continues to export sizable amounts of petroleum products illegally—and that the Iraqi government refuses to permit the U.N. to oversee or monitor these sales—strongly suggests that the proceeds from these sales are intended for non-humanitarian purposes," said Under Secretary of State Thomas R. Pickering. "We are currently seeking ways to make the Iraqi government accountable for this illegal traffic—or to end it through tougher enforcement measures." [8]

Critics say the U.S. efforts are doomed. "American policy [toward Iraq] is nothing more than the desperate embrace of sanctions of diminishing effectiveness punctuated by occasional whining, frequent bluster, political retreat and military paralysis," said Richard Perle, assistant secretary of Defense for international security during the Reagan administration. "The pressure to relax the sanctions, which has already pushed to more than $10 billion per year the amount of revenue Iraq is allowed from the sale of oil, will not subside and will almost certainly increase. The French, Russians and others will continue to agitate for the further relaxation of sanctions, and the United States will almost certainly make further concessions in this regard." [9]

U.S. policy toward Iran, whose 93 billion barrels of proven oil reserves place it fifth in the global ranking of oil powers, elicits similar concerns. Since the 1979 Islamic revolution, the United States has identified Iran as one of the world's leading supporters of international terrorism. With the Iran-Libya Sanctions Act, the United States unilaterally barred American and foreign companies from investing more than $20 million in Iran's energy sector.

But Iran is an important source of Persian Gulf oil, and several foreign companies have defied the ban. Under pressure from Europe, Clinton announced in May that he would not impose sanctions on three French, Russian and Malaysian companies that invested $2 billion in a natural gas field in Iran.

Meanwhile, U.S. oil companies are losing lucrative oil contracts to overseas competitors. "There's no question that unilateral sanctions against Iran are hurting U.S. companies," says Murphy of the API. "By preventing them from competing in those markets, the sanctions provide a relative advantage to foreign oil companies. That's a real concern of

Getting Oil Out of the Caspian . . .

Although some parts of the Caspian region and Central Asia do not hold promise as oil and gas producers, virtually every country in the vast area is likely to play a role in the industry's development. Some will serve as routes for pipelines or railroads and some, because of the threats they pose to the region's political stability, will retard progress. Here are the likely key players:

Armenia—In 1988, the largely Armenian population of Nagorno Karabakh, a province of Azerbaijan, began a six-year rebellion in an effort to become a part of neighboring Armenia. With support from Armenia, the rebels emerged victorious in 1994. Under the truce, Armenia was left in control of the province and other territory comprising 20 percent of Azerbaijan. Peace talks led by the United States, France and Russia, under the aegis of the Organization for Security and Cooperation in Europe, produced a compromise by which Nagorno Karabakh would be an autonomous province of Azerbaijan. After agreeing to these terms, rejected by the province's rebels, Armenian President Levon Ter-Petrossian was forced from office and replaced in March 1998 by Robert Kocharian, a Karabakh native who opposed the compromise. The U.S.-backed proposal to build a pipeline from Baku, Azerbaijan, to the Mediterranean port of Ceyhan, Turkey, would pass through Armenia.

Azerbaijan—Foreign oil companies have already invested heavily in the oil fields centered around the city of Baku. American companies Amoco, Exxon, Pennzoil and Unocal lead a consortium of 11 companies from eight countries—the Azerbaijan International Operating Company—that are drilling for oil for the first time in the country since it gained independence with the Soviet Union's collapse in 1991. President Heidar Aliev, a former KGB general and member of the Soviet Politburo, took power in a 1993 coup against democratically elected Abullaz Elchibey. Aliyev has survived two attempted coups since then but remains a popular leader who has brought political stability despite the destabilizing effect of the war with Armenia, which ended with the occupation of a fifth of Azerbaijan's territory, including Nagorno Karabakh.[1] Azerbaijan subsequently imposed a trade embargo against Armenia. Under pressure from the Armenian Assembly of America, Congress passed a measure in 1992 barring economic aid to the Azeri government until it takes steps to lift the embargo. An amendment to the Freedom Support Act that provides aid to the former Soviet republics, Section 907, remains in effect despite growing opposition from the Clinton administration and many lawmakers who see it as an obstacle to ensuring U.S. access to Azeri oil.

Georgia—This former Soviet republic has been torn by fighting among several distinct ethnic communities after gaining independence in 1991. President Eduard Shevernadze, a former Soviet foreign minister, has survived two assassination attempts since assuming power in 1992. An oil pipeline links the Azeri oil fields at Baku and the Georgian Black Sea port of Supsa.

Iran—A longstanding oil producer of the Persian Gulf region, Iran also borders the Caspian's southern coast. Because the region's main oil and natural gas deposits lie north of the Iranian coast, Iran's main potential role in the Caspian Sea's oil industry is as a transport link. The Caspian oil producers back construction of a pipeline that would carry the region's oil through Iran to the Persian Gulf for shipment through the Strait of Hormuz to

our members. Whether or not there's a fair foreign policy tradeoff, I can't say."

Some critics see a clear clash of interests between U.S. foreign policy in the region and U.S. energy security. "The United States' vilification of Iran, Iraq and Libya makes for strange policy," Campbell says, "especially when you understand we'll be dependent on these three places for oil before long."

The Clinton administration continues to defend its policy of containment toward Iran. "Unilateral sanctions have proven costly to U.S. business," conceded Assistant Secretary of State for Near Eastern Affairs Martin S. Indyk. "However, we believe that Iran poses threats so significant that we have no choice but accept these costs. Economic pressure has an important role in our efforts to convince Iran to cease its efforts to acquire

. . . Involves Many Actors, Many Ifs

market. The United States, which maintains a unilateral embargo against Iran for its role in supporting international terrorism, adamantly opposes this route. For now, some Caspian Sea oil is making its way indirectly through Iran via a swap arrangement by which oil is shipped to refineries in northern Iran, and an equivalent amount of Iranian oil is loaded onto tankers in the Persian Gulf for transport to market.

Kazakhstan—Though only a fraction of this vast and sparsely populated country's territory lies near the Caspian, it holds one of the region's most promising oil and gas deposits—the Tengiz Basin now under development by Chevron and other companies. China plans to build a 1,900-mile pipeline from the basin across Kazakhstan to Xinjiang, China. President Nursultan Nazarbaev, a former first secretary of the Kazakhstan Community Party, was elected after the country declared its independence from the Soviet Union in December 1991. Since then he has consolidated his power by eroding the country's limited representative government.

Kyrgyzstan—Considered to be the most democratic country in Central Asia, Kyrgyzstan is led by Askar Akaev, a former physicist who often quotes Thomas Jefferson. Elected president by the Supreme Soviet in 1990, Akaev won re-election in 1995. The country's oil potential is uncertain.

Russia—Like Iran, Russia stands to play a marginal role in the Caspian's oil production, but a crucial one in transporting the region's oil and gas to market. The first developer of the region's oil a century ago, Russia—and later the Soviet Union—largely abandoned the Caspian fields in favor of other domestic reserves to build its considerable oil industry. Today Russia transports Caspian oil by rail to the Baltic Sea and maintains other oil and gas pipelines, including one from Baku to its Black Sea port of Novorossisk. A proposed pipeline would also link the Tengiz oil field and Novorossisk.

Tajikistan—Last summer, Tajik President Imomali Rakhmonov and opposition forces signed a peace agreement ending a five-year civil war waged among the country's four regional tribes and between the secular government and Islamic militants. One of the poorest of the former Soviet republics, Tajikistan supports a thriving drug trade that has hampered efforts to improve the economy.[2]

Turkmenistan—Saparmurat Niyazov, a former Communist Party leader who became president in 1990, heads an oppressive regime based on a cult of personality. Known as Turkmenbashi—"head of the Turkmen"—Niyazov has banned opposition parties and presides over the legislature. Despite considerable reserves of natural gas and oil under and around the Caspian Sea, mismanagement of the economy has impoverished the country, which borders Afghanistan and Iran. The only existing export outlets for Turkmen gas is through Russian pipelines.

Uzbekistan—With almost 24 million inhabitants, Uzbekistan is the most populous country in the region. Cotton is the country's main product, but the government has announced plans to search for oil and gas under the polluted Aral Sea, drained of much of its water for irrigation. President Islam Karimov, a former communist leader, has introduced limited democratic reforms but faces the growing influence of Islamic militants.

[1] See Richard C. Longworth, "Boomtown Baku," *The Bulletin of the Atomic Scientists*, May/June 1998, pp. 34–38.

[2] See Martha Brill Olcott, "The Caspian's False Promise," *Foreign Policy*, summer 1998, pp. 94–113.

weapons of mass destruction and missiles and to support terrorism."[10]

The U.S. position toward Iran has shown signs of softening since the election last August of President Mohammad Khatami, who has loosened somewhat the strict regime set in place by his militant predecessors by expanding press freedoms and establishing more cordial relations with countries in the Persian Gulf and Europe.

But, citing evidence that Iran's support of terrorism is unchanged, the United States officially has not changed its position on sanctions. On June 17, however, Secretary of State Madeleine K. Albright held out the possibility of a future improvement in bilateral relations. Although she refrained from proposing specific steps to normalize relations, Albright said, "Obviously, two decades of mistrust cannot be erased overnight. The gap between us remains

Varied Sources Provide U.S. Oil Imports

U.S. oil imports come from a wide range of sources. Four of the top five providers are not Arab nations, and two of the five are not members of the Organization of Petroleum Exporting Countries (list at left). More than half the oil imported by the United States comes from non-OPEC members (list at right).

Top Five U.S. Suppliers

(thousands of barrels per day)

Venezuela*	1,657
Saudi Arabia*	1,508
Canada	1,460
Mexico	1,235
Nigeria*	812

OPEC vs. Non-OPEC Production

(thousands of barrels per day)

Arab OPEC	2,199
Other OPEC	2,519
Non-OPEC	4,976
Total	9,694

* *OPEC members*

Source: Energy Information Administration, Monthly Energy Review, *June 1998*

wide. But it is time to test the possibilities for bridging this gap." [11]

The Caspian Sea region is another area where foreign policy goals may clash with those of energy security. "Our interest in the Caspian is not defined simply by the region's energy resources, but no one doubts their significance," said Stephen Sestanovich, special adviser to the secretary of State for the new independent states. "Energy could become a source of conflict, a lever of control or an obstacle to progress. Or it could become a ticket to prosperity and peace, a secure link to the outside world." [12]

The United States is playing an active role in brokering peace talks in some of the region's simmering ethnic battles, such as those between the government of Azerbaijan and leaders of Armenian rebels who occupy the region of Nagorno Karabakh. It also has funneled economic and technical assistance—totaling $372 million in fiscal 1998—to the newly independent countries of the region, which own potentially huge deposits of oil.

But an existing measure, Section 907 of the 1992 Freedom Support Act, bars the United States from extending this assistance to Azerbaijan. Enacted at the behest of the U.S. Armenian lobby, which claimed that the government of Azerbaijan subjected ethnic Armenians to human rights abuses, the measure has attracted widespread opposition since the extent of Azerbaijan's oil wealth has become more apparent. The Clinton administration also supports the measure's repeal.

Caspian Sea oil cannot reach consumers until a pipeline network is built linking the remote, landlocked region to seaports far away. *(See map, p. 110.)* The most direct route, and possibly the least expensive to construct, would pass through Iran to the Persian Gulf. But U.S. sanctions against Iran stand in the way of that option. Instead, the United States backs a multiple pipeline network that includes the so-called main export pipeline, which would pass through Georgia and NATO ally Turkey to the Turkish port of Ceyhan on the Black Sea, where the oil would be loaded onto tankers and shipped to markets via the Mediterranean Sea.

Some experts say the United States should ease its current opposition to the Iran pipeline option. "If the international oil companies working in Central Asia do not need to start construction of new pipeline routes immediately, the U.S. government should not lock the door prematurely against the prospect of a new pipeline transiting Iran," said Richard W. Murphy, senior fellow for the Middle East at the Council on Foreign Relations. "The routing of new pipelines will have profound political and economic implications for years to come." [13]

Will the United States be ready with alternative energy sources when the oil runs out?

As a finite resource, petroleum will not last forever. Estimates of how long the world has before the oil runs out vary widely. The United States is likely to run out far sooner than many major producers, however. Using 6.8 billion barrels of oil products a year, the United States is the world's most voracious oil consumer. And it shows no sign of changing its energy habits. U.S. demand for oil is expected to grow by 20 percent by 2015. [14]

But the crunch is likely to come long before the world exhausts its oil supplies. "The idea of running out of oil, of when the last barrel comes out of the ground, is a red herring," Campbell says. "It misses the point. What's much more relevant is when world production will peak, and that will occur within the first few years of the next century." When that happens, Campbell predicts, oil consumers will be in for a major price shock as production slows.

Campbell says the United States may be especially ill-prepared for the next energy crisis. Even though U.S. oil production started declining in 1971, the United States was never seriously affected because it was able to import its oil from other countries. "This chapter is coming to an end because the other places, too, are getting close to peak production," he says. "There's monumental ignorance in the government and among the public at large on this subject."

Other critics charge that the United States, by holding oil taxes below those of other consuming nations, has encouraged oil consumption with little concern for the consequences. [15] "Part of our energy policy seems to be to keep prices low," says George Yates, chairman of the Independent Petroleum Association of America, whose 8,000 members produce almost half of the nation's domestic crude. "But we have only about 3 million barrels a day worldwide of excess capacity, and that includes Iraq, which is officially out of production because of the embargo. That's a pretty slim margin. It's like operating a factory at 100 percent capacity—you can't go on forever like this."

The development of alternatives to oil has proceeded slowly in the United States. After the energy crises of the 1970s, the federal government funded research and development of renewable energy sources such as solar, wind, geothermal and biomass energy. [16] But as oil prices subsequently fell and Congress turned its attention to reducing the federal budget deficit, support for these efforts dwindled.

According to Energy Under Secretary Ernest Moniz, his department's budgets for energy research and development fell fivefold from 1978 to 1997; privately funded research has also declined. [17] Today, much of the U.S.-developed technology for renewable energy sources is being used more intensively in Europe and Asia—where oil products are heavily taxed—than in the United States.

"The effects of world dependence on Middle Eastern oil means that while the quoted market price per barrel is about $20, the costs associated with keeping shipping lanes open, rogue states in check and terrorists at bay may more than quadruple the price per barrel," said Sen. Richard G. Lugar, R-Ind. "Given these costs, the United States may pay more than $100 billion this year for oil from the unstable Middle East. By contrast, the United States will spend less than $1 billion this year on energy research." [18]

Some analysts find little reason for concern about the United States' energy future. In their view, the technology that has enabled oil producers to discover new deposits and remove more oil from existing wells will continue to advance. "We've made tremendous advances that have made it cheaper to find and produce oil," says Murphy of the API. "If those continue, we're probably looking at oil availability for the indefinite future at today's prices or less."

When the oil crunch does come, Murphy predicts, the technology to provide cheap alternatives will be ready to take up the slack. "My feeling is that long before the use of petroleum is diminished, other sources will be there. In 10 years, all cars may be run with fuel cells instead of gasoline."

Background

OPEC's Power Play

The United States was the world's dominant oil producer for the first half of this century. [19] Outside the United States, world production fell under the control of the world's major oil companies, known as the "Seven Sisters," which acquired the right to extract oil from countries where they operated in exchange for royalties paid to the host governments. So great was their power over the markets that they were able to manipulate the price of crude from their extensive holdings in the Middle East. [20] By the late 1950s, however, they faced growing competition from independent companies and cut their prices.

Because they collected taxes based on oil prices, Persian Gulf countries where the Seven Sisters extracted oil were faced with falling revenues. On Sept. 14, 1960, representatives of Iran, Iraq, Kuwait, Saudi Arabia and Venezuela met in Baghdad, Iraq, and founded OPEC. The fledgling cartel froze oil prices to prevent further erosion in oil revenues. Other producer nations joined OPEC, including Qatar, the United Arab Emirates in the Persian Gulf region, as well as Algeria, Indonesia, Libya and Nigeria. Membership in the organization enabled these countries to set a minimum royalty to be paid by companies for the privilege of extracting oil from their

Public Broadcasting Service

A replica of the world's first oil well, drilled by "Colonel" Edwin L. Drake in Titusville, Pa., in 1859.

territories. The organization's expansion also helped the Seven Sisters by making it harder for independent companies to undercut them in the host countries.

The new order in oil development began to unravel in 1969, when Libyan strongman Muammar el-Qadaffi forced Occidental Petroleum, an independent American operator in Libya, to cut production. Because Libyan oil was of high quality, it was in high demand, and the cutback created an oil shortage. OPEC decided to profit from the change by raising oil prices. In an effort to stabilize the market, OPEC and the oil companies agreed in 1971 to a new pricing system that allowed for prices to be negotiated every five years.

Energy Crises of the 1970s

OPEC's new pricing system quickly fell apart, however. Oil was bought from companies in the open market for more than the established price, and OPEC members wanted to share in the profits. After the companies balked at their request, OPEC unilaterally raised the official price by 70 percent in October 1973, to $5.11 a barrel. The same month, the Arab producing countries imposed an oil embargo on the United States and the Netherlands for their support of Israel in the Yom Kippur War. The embargo was later replaced with a cutback in production by all Arab members except Iraq.

"Using oil supply as a political weapon was a new development in the industry and one that did great damage to OPEC's commercial credibility as a reliable supplier," writes Fadhil J. Chalabi, director of the Center for Global Energy Studies in London and acting secretary of OPEC from 1983–88. [21]

The production cutback reduced availability to all oil importers and led to a quadrupling of prices and the decade's first oil shock. The second shock came in the winter of 1978–79 following the Iranian revolution, which led to the ouster of Shah Mohammed Riza Pahlavi and his replacement by the Ayatollah Ruhollah Khomeini's militant Islamic regime.

The revolution caused a disruption of oil from the Persian Gulf that was compounded by the outbreak in 1980 of the Iran-Iraq War. By January 1981, OPEC's oil price had reached $34 a barrel, more than 10 times the price in 1972, before the first oil shock. Taken together, the shocks produced a windfall for OPEC members, whose oil revenues skyrocketed from less than $23 billion in 1972 to more than $280 billion by the end of the decade.

Oil Consumers React

But OPEC's bonanza days were numbered. The United States and other industrial countries reacted strongly to the gasoline rationing, long lines at the pump and double-digit inflation produced by the production cutbacks and price increases. They launched a frantic search for alternative sources of oil and set about trying to reduce their dependence on oil imports by making more fuel-efficient cars, improving energy conservation and developing renewable energy sources.

Over time, the growth in demand for oil slowed. By 1995, industrial nations were consuming only 2 million barrels a day more than they had in 1975. "Put another way," writes Chalabi, "oil consumption by OECD countries in 1996 was less than at its peak level in 1978, even though their [gross domestic product] had grown by 42 percent during the same period." [22]

The oil consumers' efforts to reduce their reliance on Middle East oil paid off most successfully in the development of new oil fields. Companies shifted their investments to non-OPEC countries such as Mexico and Canada, as well as Britain and Norway, where they developed new platform-drilling technology to exploit the vast deposits under the North Sea. High oil prices also made it feasible to tap the enormous reserves on Alaska's North Slope.

As a result of these efforts, OPEC's share of the global oil market fell by half, from 56 percent in 1975 to just 26 percent in 1995. Although non-OPEC countries possess only a quarter of the world's oil reserves, they now account for 60 percent of global production.

Current Situation

OPEC's Dilemma

Despite the industrial nations' success in reducing their dependence on Middle East oil, OPEC has continued to reap enormous profits from oil exports. The revenues from oil sales have enabled many member countries to invest in other industries to help diversify their economies in preparation for the day when they will no longer be able to rely on petroleum exports.

But the glut in oil supplies that began last fall has dealt a serious blow to countries that still depend heavily on oil exports, both within and outside OPEC. Even Saudi Arabia is feeling the pinch of falling oil revenues.

The economic crisis has encouraged cheating on the part of some OPEC members, further undermining the organization's clout. Venezuela is said to be the most flagrant offender, and has called on fellow OPEC members to abandon quotas altogether in favor of other strategies to win market share. Algeria, Iran, Libya and Nigeria also routinely ignore the production quotas to prevent further erosion of the oil revenues they depend on for economic survival. [23]

In a desperate effort to slow production, OPEC for the first time called on non-member producer countries to cooperate with its quota system. In March, Mexico agreed to cut oil output, and Norway later agreed to curb production as well. Russia has expressed interest in joining OPEC and attended the organization's June meeting in Geneva as an observer. The March agreement sent oil prices up by 13 percent, to almost $17 a barrel. But prices have since dropped back to around $14, suggesting that the market is not confident that OPEC can hold production down for long.

Some U.S. oil-company representatives support OPEC's campaign to hold the line on output and prices. "I'm hoping that by including non-members, OPEC can exercise some restraint on oil production," says Yates of the Independent Petroleum Association. "Restraint, coupled with higher prices, will mean fewer oil wells being abandoned in the United States. It also means new wells will be drilled to find additional oil, which will make us better able to meet demand in the future. OPEC's really the American consumer's friend right now."

But many analysts question OPEC's ability over the long term to control the global market, which has changed in fundamental ways over the past two decades. "What's dominating the market today is the advance of technology, which is having a significant impact on the industry's ability to explore and produce oil in areas they were forbidden from exploring before," says Murphy of the API. Another change is the willingness of countries such as Venezuela, where the oil business was once run by state-owned companies, to have foreign, private companies come in and produce their oil. Finally, the former Soviet republics in the Caspian may have a major impact on world markets when the region's oil starts to flow. "OPEC is a bystander in this region," Murphy says.

OPEC's difficulties lead some experts to predict its eventual demise. "OPEC, as such, is disintegrating," Campbell says. Indeed, OPEC members know that drastically raising prices to solve their current dilemma will backfire on them in the end. "They remember what happened after the last shock," he says, "and they fear that if they put up the price they will lose their market share, and if that happens that they will lose everything."

In Campbell's view, however, the five leading Middle Eastern producers—Iran, Iraq, Kuwait, Saudi Arabia and Abu Dhabi (one of seven emirates comprising the United Arab Emirates)—have nothing to worry about. "They don't realize that the situation is very different today," he says. In about four years, by his calculation, world production of conventional oil—that which is easily recoverable—will peak, sending prices upward once again.

"Unlike the 1970s, when a flood of new oil production followed the price shocks," he says, "there are very few new oil deposits being found, with the exception of the Caspian." Campbell predicts that Middle Eastern producers, with about half the world's remaining conventional oil reserves, will see their share of recoverable oil rise substantially by 2000.

At that point, he says, "the Middle Eastern countries will recognize their control."

Cheap Oil

For now, however, oil-producing countries, including the petroleum-rich kingdoms of the Middle East, are suffering as a result of low oil prices. Most depend on oil exports for the bulk of their revenue. Saudi Arabia has seen its oil revenues drop from $43 billion last year to an estimated $29 billion in 1998. As a result, King Fahd's government has been forced to borrow $2 billion from Saudi banks to fund public programs and has cut its budget by at least 10 percent. [24]

As Oil Runs Out, Technology Buys Time

The world's supply of "conventional" oil—oil that is easily recovered—is running short of demand. In little more than a decade, some experts predict, global demand will so far exceed supplies of conventional oil that price shocks will occur that may lead to recession or political turmoil.

Faced with this impending shortfall, oil companies are investing heavily in research to improve existing technologies and develop new ones. Some of the work is already paying off. According to Roger N. Anderson, director of petroleum technology research at Columbia University, recent advances in finding and extracting oil may raise world oil production by more than 20 percent by 2010. [1]

Recent advances in oil exploration and extraction include:

4-D Seismic Analysis—As oil and natural gas are extracted from underground deposits, the remaining oil and gas seeps into the layers of rock. Three-dimensional monitoring with seismic instruments helps identify the location of oil deposits but cannot follow the shifting of oil that occurs as the well's contents run down. Anderson and others have developed a "4-D" system that incorporates the added dimension of time and helps drillers determine where the rest of the oil is likely to settle. Recovering this otherwise lost oil can increase the output of a given field by as much as 15 percent. The new technology has been applied at about 60 oil fields worldwide over the past four years.

Steam and Gas Injection—Drillers traditionally abandon wells when the flow of oil slows to a trickle. But scientists now know that this often leaves behind more than half the oil in a given deposit. Pumping steam, natural gas or liquid carbon dioxide into seemingly dry wells can force the remaining oil through porous rock toward a neighboring well, where it can be extracted. Another technique involves pumping water below the deposit, which increases the pressure under the oil, forcing it to the surface. Although steam or gas injection increases oil recovery by up to 15 percent, the high cost of this technique often outweighs the oil's value.

Directional Drilling—Oil wells typically are drilled straight down into the ground. But new technology allows drillers to change direction thousands of meters below ground and bore horizontally through rock in search of deposits a mile or more away from the wellhead. Sensors near the drill bit can detect oil, water and gas by measuring the density of surrounding rock or by measuring minute changes in electrical resistance. Engineers at the surface monitor the drill's progress by computer.

Deep-Water Drilling—Most offshore rigs, such as those along the coasts of Texas and Louisiana, operate at relatively shallow depths. But new technology is enabling drillers to tap into oil deposits under deeper water—currently down to 1,700 meters. Unmanned submarines install equipment on the ocean

The bad news for oil producers may have a silver lining over the longer term. As autocratic regimes in the Middle East are forced to reduce their generous health, education and welfare programs, social unrest may quicken the pace of economic and political reforms. Pressure for reform has surfaced in a number of the region's oil states, including Saudi Arabia and the other gulf states, as well as Syria and Iran. [25]

Oil producers outside the Middle East are suffering as well. Mexico, a major source of U.S. oil imports, announced its third budget cut in six months in early July, citing unexpectedly low oil revenues. Nigeria, in the midst of political turmoil in the wake of the sudden death of popular opposition leader Mashood Abiola, faces an even graver plight. In this poor West African country, earnings from oil exports provide a vital buffer against widespread poverty. For Russia, the price drop has only added to the country's serious economic crisis. Only Britain and Norway, with the most diversified industrial economies among the major foreign exporters, can absorb the loss of oil revenues without major disruption.

In the United States, which now imports more than half its oil, the fall in prices is a mixed bag. American

floor to regulate the flow of oil at high pressure and prevent environmentally devastating blowouts. The oil is then loaded onto tankers at sea or piped ashore or to shallow-water platforms through underwater pipelines. Recently declassified U.S. Navy technology enables geophysicists to detect underwater oil deposits through the sheets of salt and basalt that often hide them from conventional seismic surveys. While deep offshore drilling is very expensive, it is expected to become more widely pursued as conventional oil deposits dry up. Deep-water platforms are already in use off the coast of Newfoundland, Canada, and more are planned for the Gulf of Mexico, the North Sea and the Atlantic Ocean off Brazil and West Africa.

Just how much impact technological advances can have on global oil supplies is a matter of heated debate. Oil companies are optimistic about technology's ability to extend the petroleum era for decades to come. "If you look at the available proved reserves, there are about 1 trillion barrels of oil still in the ground," says Edward H. Murphy, an economist at the American Petroleum Institute. "If we continue to produce 27 billion barrels a year as we do now, that means we have 37 years left." That deadline can be extended, Murphy says, by technological advances. "We only recover 40 percent of production out of a given field today," he says. "So there's substantial room for enhanced recovery."

There's a downside to technology's impact on oil production, however, especially for small domestic producers who depend on the slow but steady flow of oil from marginal wells. "Some technologies mean that the same reserves are depleted more quickly instead of maintaining production over a long period," says George Yates, chairman of the Independent Petroleum Association of America. "So while technology has had a very positive impact on this industry, it's also put more oil on the market, which exacerbates our problems."

Still other experts say technology can only delay for a short time the inevitable demise of our oil-based economy. "Deep-water drilling is capable of producing 100 billion barrels, or about five years of world demand," says Colin J. Campbell, a consultant in Geneva, Switzerland. "It's expensive, and it's viable only in giant fields." He foresees further development of oil deposits under polar ice in Alaska, Russia and Canada, as well as the large deposits of heavy oil in Canada and Venezuela. "But even that won't make a lot of difference," Campbell says. "Whatever the accumulated advances of technology can deliver ought to be incorporated into our estimates of existing oil reserves. It's not something you can keep adding to."

[1] Information in this section is largely based on Roger N. Anderson, "Oil Production in the 21st Century," *Scientific American,* March 1998, pp. 86–91.

motorists are benefiting from cheap gasoline, and low prices are helping keep inflation in check. But the U.S. oil industry is facing the same problems as the major producing nations. Occidental Petroleum reported a 66 percent drop in net income for the second quarter of 1998, and other major companies are expected to issues similarly bad news. [26]

Lower crude oil prices have dampened the incentive for oil companies to look for new domestic deposits and to maintain production in existing fields with marginal output. According to the API's Murphy, the number of drilling rigs used to search for oil and gas has fallen by 31 percent over the average used for that purpose over the past decade. "This drilling data should, we believe, be of major concern to those interested in this country's energy future," he said. "Oil drilling at these rates is inadequate to maintain U.S. crude oil production levels, particularly in the lower 48 states." [27]

Many smaller American oil companies are in especially bad shape. As a group, the 8,000 independent oil companies suffered a 25 percent drop in revenues over the past year. These include many companies that extract natural

Part of the Caspian Sea oil-drilling installation built by Soviet leader Josef Stalin in 1949, about 40 miles east of Baku, Azerbaijan.

AP/Shakh Aivazov

gas as well as oil. Natural gas prices have been largely unaffected by the oil glut. "But some of our members produce only oil," explains Yates of the Independent Petroleum Association. "For them, the impact of low oil prices is extremely dramatic. You have to go back to the 1930s for corresponding prices, and then they had to call out the Texas Rangers to bring order to the market." *

Many small companies face bankruptcy. "The worst off are those with marginal production," Yates says. "There are thousands of these mom-and-pop operations." As large oil fields have run dry in the United States, many of the 500,000 wells operated by the association's members produce just two or three barrels of crude a day. "Most of these marginal wells could go on producing forever, but not at a loss, so they are being abandoned," Yates says. "This is a very serious issue, because the oil we don't produce domestically we have to import. And it's not just the oil producers who are affected, because as imports grow, so does the trade deficit." In May, the U.S. trade deficit hit a record $15.7 billion as imports grew and exports fell, especially to the economically troubled Asian countries. [28]

*In 1931, an oil glut in Texas and Oklahoma drove down oil prices, prompting Texas Gov. Ross Sterling to send the Texas Rangers into the East Texas oil fields to enforce a production cutback.

Caspian Treasure

As petroleum deposits are depleted in coming years, prices are likely to rise, ending the era of cheap oil. With few major new fields expected to be found, oil companies are in hot competition to develop the potentially huge reserves under and around the landlocked Caspian Sea, which lies between southern Russia and northern Iran. Estimates of the volume of crude oil in the region range up to 200 billion barrels—more than a quarter of the Middle East's reserves.

The region's oil has been known about for centuries. Marco Polo remarked on the seepage of oil around the Caspian Sea during his travels along the Silk Road, the ancient trade route between Europe and the Far East. Swedish businessmen Ludwig and Robert Nobel began developing oil fields in Baku, on the western coast of the Caspian Sea, more than 120 years ago. [29] The Soviet Union, with large oil reserves in other parts of the country, did little to develop the Caspian fields because most of the oil was trapped under water or salt formations and too hard to extract.

Modern technology now places the Caspian's oil within reach, and oil companies have descended on the sparsely populated region since 1991 to gain a foothold in what may be the world's last oil boom. One of the first Western companies in the region was Chevron Corp., which in 1993 began investing in Kazakhstan's Tengiz oil field in and around the northeastern Caspian. Numerous

©Digital Stock

Oil companies are investing heavily in research to improve exploration and drilling technology, because conventional methods may be unable to meet demand in the near future.

other companies have since invested in the region as well, though Chevron has spent more than all of them put together. [30]

The main obstacle to the development of Caspian oil is the volatile political situation throughout the region. The Soviet Union's dissolution in 1991 transformed the former Soviet republics of Azerbaijan, Kazakhstan and Turkmenistan, which also border the Caspian, into independent countries. With no democratic tradition, ethnic divisions that rival those of the former Yugoslavia and virtually undeveloped economies, the Caspian is a region that journalist Richard C. Longworth describes as "Bosnia with oil."

"In this oil-soaked cockpit, the prospects for both wealth and trouble are simply stupendous," he writes. "The Caucasus is a land of ancient vendettas and warring tribes that makes the Balkans look straightforward by comparison, and the Caspian is where this bloody region meets Central Asia and the Middle East. It's where Orthodox Christianity meets both Sunni and Shi'ite Islam, where Iranians traveling north have met Russians coming south. It's an area once ruled by Iran, then by Russia, now contested by Turkey, whose language and civilization dominate the region." [31]

Pipelines and Politics

The Caspian region's political turmoil exacerbates another significant obstacle to developing Caspian oil: transportation problems. Once the crude is extracted, it must travel thousands of miles to reach ports where it can be loaded onto tankers and shipped to markets. Some oil from the Tengiz field already reaches the Baltic Sea by rail. But rail shipment is expensive and increases the risk of oil spills. [32] The only alternative under consideration is to build pipelines. Three are already being developed. The problem is where to build the next ones.

China plans a 1,900-mile pipeline from Tengiz across Kazakhstan into China. But all the other routes from the Caspian to Western markets have political drawbacks. The most direct route to open sea, and the one desired by the Caspian countries, is through Iran to the Persian Gulf. But the United States opposes that route because it would defy U.S. sanctions against Iran and make Caspian oil hostage to political instability in the gulf.

Although Secretary Albright waived the sanctions against three foreign companies that invested in Iranian oil fields in May, the move apparently did not signal an immediate change in policy. "We continue to oppose trans-Iran pipelines for Caspian energy exports in the strongest terms," said Sestanovich of the State Department. [33]

Russia has proposed a route that would carry oil from the eastern Caspian to the Russian port of Novorossisk on the Black Sea, which already serves as the terminus for an existing pipeline from Baku. Turkey opposes this and other new pipelines that would terminate at the Black Sea because the oil must then be shipped through the narrow Bosporus strait, which passes through Istanbul, to reach the Mediterranean. Some 4,000 tankers already pass through the 17-mile-long passage each year, negotiating four 45-degree turns on their way and posing the risk of

©Digital Stock

New pipelines are needed to move crude oil from the landlocked Caspian Sea region to deep-water ports where it can be loaded onto tankers and shipped to markets.

disastrous oil spills. Another planned pipeline, linking Turkmenistan to Pakistan, is on hold pending resolution of Afghanistan's protracted civil war.

The Clinton administration supports a network of multiple oil and gas pipelines to make it less likely that a supply interruption in one would cut off the entire region's oil flow. It is pushing strongly for a pipeline stretching from the Caspian port of Baku, Azerbaijan, through Georgia to the Mediterranean port of Ceyhan in Turkey, a NATO ally.

The region's political turmoil leads some observers to question the ability of the Caspian region to meet the growing global demand for oil. "The Caspian states are not necessarily plunging into a maelstrom in which corrupt regimes will be challenged either by secret drug lords or social unrest," writes Martha Brill Olcott, a senior associate at the Carnegie Endowment for International Peace and professor of political science at Colgate University. "But the possibility of such chaos cannot be precluded. Each of these states faces difficult political transitions in the next five to 10 years, while peak oil production and the economic benefits that it promises are unlikely to be realized in the region until 2010." [34]

Other obstacles to the region's ascendance as a leading oil provider spring from its geological limitations. "The Caspian is the only place in the world right now with significant promise," Campbell says. "But there's been some exaggerated talk about its real potential." He estimates that the region may contain only 50 billion barrels. "Even that is stretching credibility a little bit," he says. "And that's if it comes in, which as of today is not something to count on."

With Caspian deposits perhaps equivalent in volume to those under the North Sea, Campbell says, offshore production could total up to 4 million barrels a day by 2025. "Although that's valuable and not to be dismissed," he says, "it's unlikely to make much of an impact on the peak of global oil production," which he says will occur within the next few years.

Outlook

Will Prices Rise?

Despite OPEC's recent efforts to curb production, it may be some time before the current oil glut subsides enough to push prices back up. But simply stopping the current overproduction may not be enough to return global supply and demand to balance, according to the Paris-based International Energy Agency. Oil importers will first have to absorb the excess oil that is already in the supply chain. They also will have to reduce their record high levels of oil stocks, the result of a warm winter in North America and of stagnant industrial production in Japan and other East Asian countries. [35]

Some economists predict that oil prices will remain low, even after the current glut has vanished, because the world has already begun to wean itself from petroleum products.

"The industrial world uses 42 percent less oil to produce an extra unit of [gross domestic product] than it needed in 1973," writes Lester C. Thurow, an economics professor at the Massachusetts Institute of Technology. "Transportation still depends upon oil, but fuel cells look as if they are about to arrive. When they do, early in the next century, oil demand will begin to fall even in this, its primary market." As a result, Thurow predicts, oil "prices will be low for the foreseeable future." [36]

Other experts are equally convinced that oil prices are headed in the opposite direction. In Campbell's view, the

FOR MORE INFORMATION

Energy Information Administration, U.S. Department of Energy, 1000 Independence Ave. S.W., #2G051, Washington, D.C. 20585; (202) 586-5214; www.eia.doe.gov. The EIA collects and publishes data on domestic production, imports, distribution and prices of crude oil and refined petroleum products.

American Petroleum Institute, 1220 L St. N.W., Washington, D.C. 20005; (202) 682-8042; www.api.org. This membership organization of U.S. producers, refiners, marketers and transporters of oil and related products provides information on the industry.

Independent Petroleum Association of America, 1101 16th St. N.W., 2nd Floor, Washington, D.C. 20036; (202) 857-4722; www.ipaa.org. Members are independent oil and gas producers and others involved in domestic oil and gas production.

International Energy Agency, 9, rue de la Federation, 75739 Paris Cedex 15, France; (33-1) 40.57.65.54; www.iea.org. Created in the wake of the energy crises of the 1970s, the IEA monitors global oil production and helps consumer countries coordinate strategies to avoid supply disruptions.

world is in for an oil shock that will make the energy crises of the 1970s pale in comparison. "There will be an initial price shock around 2000, with a doubling or perhaps a tripling of prices," he predicts. "Although the roof won't fall in overnight, long-term shortages will force a change in attitude about energy consumption."

The price rise will spur development of renewable energy sources, Campbell predicts. "But it's hard to picture this being done at a rate and a scale to enable renewables to act as substitutes for the way we've used cheap oil up to now."

In Campbell's view, the United States will play a crucial role in determining the global response to what he sees as the coming oil crisis. "There is an enormous danger that the United States, with its peculiar Middle East policies, may misunderstand this situation and perceive the price hikes to be a politically hostile act by Iran or Iraq," he says.

"But at the heart of the matter, the coming oil crisis isn't about politics; its simply about the distribution of a resource that events during the Jurassic Period dictated." ❖

Thought Questions

1. Senator Richard Lugar noted (p. 117 of this article) "The effects of world dependence on Middle Eastern oil means that while the quoted market price per barrel is about $20, the costs associated with keeping shipping lanes open, rogue states in check and terrorists at bay may more than quadruple the price per barrel. Given these costs, the United States may pay more than $100 billion this year for oil from the unstable Middle East. By contrast, the United States will spend less than $1 billion this year on energy research." How do you think public perception about oil might change if these hidden costs were more widely discussed? Does the relatively low level of spending on energy research surprise you?
2. Is the price of military actions to ensure a steady supply of oil to the United States a necessary and valid expenditure of taxpayer money or a corporate subsidy that keeps the price at the pump artificially low?
3. If pump prices were to reflect the actual cost of oil production and transport to the United States, what impact might that have on the average consumer's attitudes toward oil consumption?
4. If you were an urban planner, what steps could you take to ensure minimization of use of fossil fuels in your community?

Notes

[1] Peña testified April 30, 1998, before the House International Relations Committee. Peña announced on April 6 he would resign this summer, and President Clinton nominated U.N. Representative Bill Richardson as the next Energy secretary. The Senate Energy and Natural Resources Committee voted 18–0 on July 29 to approve Richardson's nomination. The Senate approved Richardson on July 31.

[2] For background, see Mary H. Cooper, "Oil Imports," *The CQ Researcher,* Aug. 23, 1991, pp. 585–608.

[3] For background, see Rodman D. Griffin, "Mexico's Emergence," *The CQ Researcher,* July 19, 1991, pp. 497–520.

[4] For background, see David Masci, "U.S.-Russian Relations," *The CQ Researcher*, May 22, 1998, pp. 457–480, and Mary H. Cooper, "Russia's Political Future, *The CQ Researcher,* May 3, 1996, pp. 385–408.

[5] For background, see Patrick G. Marshall, "Calculating the Costs of the

Gulf War," *Editorial Research Reports,* March 15, 1991, pp. 145–156.

[6] See James Kim and Chris Woodyard, "Glut Knocks Oil Costs Down, but It Won't Last," *USA Today,* Feb. 26, 1998.

[7] See "Saddam Seeks End to U.N. Embargo This Year," *The Washington Post,* July 18, 1998.

[8] Pickering testified May 21, 1998, before a joint hearing of the Senate Energy and Natural Resources and Foreign Relations committees.

[9] Perle testified at the May 21 hearing before the Senate Energy and Foreign Relations committees.

[10] Indyk testified May 14, 1998, before the Senate Foreign Relations Subcommittee on Near Eastern and South Asian Affairs.

[11] Albright spoke before the Asia Society in New York City. See Thomas W. Lippman, "Albright Offers Iran Possibility of Normal Ties," *The Washington Post,* June 18, 1998.

[12] Sestanovich testified July 8, 1998, before the Senate Foreign Relations Subcommittee on International Economic Policy, Export and Trade Promotion.

[13] Murphy testified May 14, 1998, before the Senate Foreign Relations Subcommittee on Near Eastern and South Asian Affairs.

[14] See Tad Szulc, "Will We Run Out of Gas?" *Parade Magazine,* July 19, 1998, pp. 4–6.

[15] For background, see Mary H. Cooper, "Transportation Policy," *The CQ Researcher,* July 4, 1997, pp. 577–600.

[16] For background, see Mary H. Cooper, "Renewable Energy," *The CQ Researcher,* Nov. 7, 1997, pp. 961–984.

[17] Moniz testified Feb. 5, 1998, before the House Commerce Subcommittee on Energy and Power.

[18] Lugar, chairman of the Senate Agriculture, Nutrition and Forestry Committee, spoke before the committee on Nov. 13, 1997.

[19] For background, see Cooper, *ibid.*

[20] Unless otherwise noted, information in this section is based on Fadhil J. Chalabi, "OPEC: An Obituary," *Foreign Policy,* winter 1997–98, pp. 126–140.

[21] Chalabi, *op. cit.,* p. 130.

[22] *Ibid,* p. 133. The Organization for Economic Cooperation and Development represents the leading industrial nations.

[23] *Ibid,* p. 136.

[24] See Youssef M. Ibrahim, "Falling Oil Prices Pinch Several Producing Nations," *The New York Times,* June 23, 1998.

[25] See "When Gulf States Tighten Their Belts," *The Economist,* March 14, 1998, pp. 49–50.

[26] See Bill Meyers, "As Oil Prices Slip, So Do Earnings: 'It's a Real Mess Out There' for Now," *USA Today,* July 21, 1998.

[27] Murphy spoke July 15, 1998, at an American Petroleum Institute press conference in Washington.

[28] See John M. Berry, "Trade Deficit Soared in May," *The Washington Post,* July 18, 1998.

[29] For background on early oil development in the Caspian region, see Daniel Yergin, *The Prize* (1991), pp. 56–65. Another Nobel brother, Alfred, invented dynamite. Their father, Immanuel, invented the underwater mine.

[30] See Craig Mellow, "Big Oil's Pipe Dream," *Fortune,* March 2, 1998, pp. 158–164.

[31] See Richard C. Longworth, "Boomtown Baku," *The Bulletin of the Atomic Scientists,* May/June 1998, p. 35.

[32] For background, see Mary H. Cooper, "Oil Spills," *The CQ Researcher,* Jan. 17, 1992, pp. 25–48.

[33] From July 8, 1998, testimony before the Senate Foreign Relations Subcommittee on International Economic Policy, Export and Trade Promotion.

[34] Martha Brill Olcott, "The Caspian's False Promise," *Foreign Policy,* summer 1998, p. 110.

[35] International Energy Agency, *Oil Market Report,* July 9, 1998. The agency was set up in the wake of the 1970s' energy crises to monitor the oil market and help correct balances in supply and demand.

[36] Lester C. Thurow, "Oil Prices No longer Hold Us Hostage," *USA Today,* May 26, 1998.

Chronology

1960s *Oil production comes under the growing control of Middle Eastern producers.*

Sept. 14, 1960
Iran, Iraq, Kuwait, Saudi Arabia and Venezuela form the Organization of Petroleum Exporting Countries (OPEC).

1969
Libyan strongman Muammar el-Qadaffi forces Occidental Petroleum, a U.S. company, to curtail production of Libyan oil, producing a shortage in world oil supplies and prompting OPEC to raise oil prices.

1970s *An Arab embargo leads to oil crises that quadruple the price of oil.*

1970
Domestic oil production peaks at 11.3 million barrels a day in the United States, forcing it to gradually increase its dependence on imports.

October 1973
After OPEC raises oil prices by 70 percent, to $5.11 a barrel, Arab producers impose an embargo on oil exports to the United States and the Netherlands for their support of Israel in the Yom Kippur War. Oil prices soar above $17 a barrel.

1975
The Strategic Petroleum Reserve is created to protect the United States from interruptions in oil supplies. Congress sets fuel-efficiency standards for cars.

December 1978
The Iranian revolution disrupts Persian Gulf oil supplies, causing a second oil shock and deepening inflation in oil-consuming countries.

June 1979
OPEC raises the price of crude from $14.50 to as high as $23.50 a barrel. Gas lines form in the United States.

1980s *As oil production spreads outside OPEC, prices begin to fall.*

1980
The eight-year Iran-Iraq War begins, compounding the disruption in Gulf oil. By the following January, OPEC's oil price reaches $34 a barrel, more than 10 times the price in 1972.

1986
Oil prices fall to their lowest level since the first oil crisis in 1973.

1990s *OPEC's inability to control output leads to a global oil glut.*

Aug. 2, 1990
Iraq occupies Kuwait, cutting off 1.6 million barrels of oil a day from the world market. The U.N. imposes an embargo on Iraqi oil exports as well. Panic buying pushes oil prices up from $13 a barrel to $40.

1991
The United States leads a coalition of forces in the Persian Gulf War to drive Iraqi occupying forces out of Kuwait.

December 1991
The Soviet Union dissolves, and the newly independent countries of the Caspian Sea begin opening their oil reserves to exploitation.

1992
Congress passes the Freedom Support Act providing economic assistance to the former Soviet republics. Section 907 of the law prohibits the aid from going to Azerbaijan—a major oil producer—for abuses against ethnic Armenians.

1993
Chevron Corp. invests in Kazakhstan's vast Tengiz oil field, beginning the oil rush in the Caspian Sea region.

1996
The U.S. begins to import more than half the oil it consumes.

December 1996
An "oil-for-food" provision is added to the U.N. embargo against Iraq, allowing the country to export just enough oil to pay for food and other essential supplies. The same year, energy conservation pushes oil consumption in the industrial world below the peak level of 1978.

October 1997
The financial crisis in Asia curbs oil consumption in that part of the world, leading to a glut in world oil supplies.

March 1998
As oil prices plummet to their lowest levels in decades, OPEC reaches an unprecedented agreement with a non-OPEC oil producer—Mexico—to curtail production in an effort to keep oil prices from falling further.

At Issue:

Should the United States ease its sanctions against Iran to improve international access to Caspian Sea oil?

Yes

S. FREDERICK STARR—Chairman, Central Asia Institute, Nitze School of Advanced International Studies, Johns Hopkins University

From testimony before the House International Relations Subcommittee on Asia and the Pacific, Feb. 12, 1998.

Two presidential directives in 1995 and the Iran-Libya Sanctions Act of 1996 cut off all significant American and foreign investment in Iran's petroleum industry, including pipelines. The purpose was to pressure Iran into dropping its support for terrorism, abandoning programs to develop atomic weapons and [stopping its] meddling in the Middle East peace process. However laudable the aims, the burden of these measures falls disproportionately on Azerbaijan, Kazakhstan and Turkmenistan, for it prevents them from exporting their gas and oil by one of the obvious alternative routes to Russia, namely, Iran. The U.S. position has been to argue that this would not be in the Central Asians' own interest, but none of our friends there agree.

Now, let us suppose that the U.S. sanctions [remained] in place for a long time and [were] truly effective. Over time, so we have argued, planners and financial markets would adjust to this reality. They would construct the east-west pipeline and thus give Central Asians access to secure export routes bypassing both Iran and Russia.

But this is not happening. French, Indonesian and Russian firms are already investing in the construction of oil facilities and pipelines in Iran, and the U.S. seems disinclined to intervene against them. Iran itself is busy constructing a line linking Turkmenistan and Turkey. Turkmenistan and Kazakhstan have worked out swap deals with Tehran, by which Central Asia ships its crude oil to Iran's north and Iran then exports the same quantity of its own oil from the south. In short, the American quarantine of 1995–6 is not holding. . . .

[The United States could] adopt a "wait and see" posture toward Iran, one that would be cautious but less categorical than our current policy. It would replace an "all or nothing" approach with one that recognizes the existence of a large number of finely calibrated positions between these two extremes. . . . On balance, it seems to me that [this alternative] holds the most promise for achieving a balance between U.S. objectives in Central Asia, in the Caspian basin and in Iran. . . .

[I]t is no longer possible to treat U.S. policy toward Central Asia and toward Iran as totally separate from one another. Our Iranian policy, however just its goals, has a powerful and, for the most part, negative impact on our ability to achieve our stated objectives in Central Asia and the Caspian basin.

No

SEN. SAM BROWNBACK, R-KAN.

From testimony before the Senate Foreign Relations Subcommittee on International Economic Policy, Export and Trade Promotion, Oct. 23, 1997.

The countries of the South Caucasus and Central Asia—Armenia, Azerbaijan, Georgia, Kazakhstan, Kyrgyzstan, Tajikistan, Turkmenistan and Uzbekistan—are at a historic crossroads in their history: They are independent, they are at the juncture of many of today's major world forces, they are rich in natural resources and they are looking to the United States for support. . . .

First of all, these countries are a major force in containing the spread northward of anti-Western Iranian extremism. Though Iranian activity in the region has been less blatant than elsewhere in the world, they are working very hard to bring the region into their sphere of influence and economic control.

Secondly, the Caspian Sea basin contains proven oil and gas reserves which, potentially, could rank third in the world after the Middle East and Russia and exceed $4 trillion in value. Investment in this region could ultimately reduce U.S. dependence on oil imports from the volatile Persian Gulf and could provide regional supplies as an alternative to Iranian sources. . . .

The independence of the region could indeed well depend on the successful construction of pipelines on an east-west axis through non-Russian as well as non-Iranian territory. Both Russian and Iranian rhetoric on this issue shows clearly that these countries see the connection between pipelines free of Russian and Iranian control and their domination over the region. And it is no coincidence that we are seeing an intense rapprochement between these two countries.

Time is of an essence here. We have the opportunity to help these countries rebuild themselves from the ground up and to encourage them to continue their strong independent stances, especially in relation to Iran and the spread of extremist, anti-Western fundamentalism, which is one of the most clear and present dangers facing the United States today.

The window of opportunity has been closed even further by the recent investment by the French company Total . . . in the Iranian South Pars offshore gas field. It is vital that the [Clinton] administration hold strong on implementing existing sanctions and on discouraging our allies from following the despicable example of Total. If the floodgates open through Iran, the eastern Caspian will certainly fall into the Eastern sphere of dominance, and the South Caucasus will lose out on its opportunity to prosper as producer of oil and as a pivotal transit point from East to West.

Bibliography

Selected Sources Used

Books

Adelman, M. A., *The Genie Out of the Bottle: World Oil Since 1970*, MIT Press, 1995.
The author, a leading petroleum economist, builds on his earlier analyses of the global oil trade with this review of events encompassing OPEC's rise to power over oil production and its more recent decline.

Campbell, Colin J., *The Coming Oil Crisis*, Multi-Science Publishing and Petroconsultants, 1997.
A former geologist for Texaco and Amoco predicts that global oil production will peak within the next few years. It is at this point, he writes, not when supplies are close to exhaustion, that oil prices will rise significantly.

Yergin, Daniel, *The Prize: The Epic Quest for Oil, Money & Power*, Simon & Schuster, 1991.
This sweeping history of the oil industry takes the reader from the first oil well in 1859 in Titusville, Pa., through the rise of OPEC and the West's reaction to the energy crises of the 1970s.

Articles

"Asia: The Cloning of America," *Energy Investor*, June/July 1998, pp. 2–3.
Despite its current economic crisis, Asia remains a leading consumer of oil. With rapid industrialization in much of the continent, Asia already uses 70 percent of all newly discovered oil, and its demand for oil can only be expected to grow in coming decades.

Chalabi, Fadhil J., "OPEC: An Obituary," *Foreign Policy*, winter 1997–98, pp. 126–140.
A former OPEC official writes that the formerly omnipotent cartel will continue to lose its control over global oil production and prices because oil-consuming countries have found alternative sources of oil. To survive, the organization's members must introduce economic reforms and change its quota system to reflect changes in the global oil market.

Coy, Peter, Gary McWilliams and John Rossant, "The New Economics of Oil," *Business Week*, Nov. 3, 1997, pp. 140–144.
The authors conclude that today's low oil prices may continue for decades as technological advances make it easier than ever to produce oil.

Longworth, Richard C., "Boomtown Baku," *The Bulletin of the Atomic Scientists*, May/June 1998, pp. 34–38.
Political turmoil in the oil-rich countries bordering the Caspian Sea makes the region the equivalent of "Bosnia with oil," writes journalist Longworth. U.S. policies, especially those favoring Armenian separatists in Azerbaijan, work against U.S. oil companies' efforts to develop the region's oil.

Olcott, Martha Brill, "The Caspian's False Promise," *Foreign Policy*, summer 1998, pp. 94–113.
A Colgate University professor of political science describes the obstacles to oil development in the Caspian Sea region, the most promising new source of oil today. Poverty, ethnic rivalries and economic mismanagement since the region gained independence with the Soviet Union's collapse in 1991 may derail plans for large-scale oil exports.

"Preventing the Next Oil Crunch," *Scientific American*, March 1998, pp. 77–95.
Four articles describe the problems associated with falling oil reserves. Technological advances will stretch out the petroleum age, but price hikes are likely as oil reserves drop and extraction becomes increasingly costly.

"When Gulf States Tighten Their Belts," *The Economist*, March 14, 1998, pp. 49–50.
The collapse in oil prices since early this year has drastically curtailed revenues in the rich kingdoms of the Persian Gulf. Because they depend so heavily on income from oil exports, these countries are having to reduce spending on social programs and hasten economic reforms.

Reports and Studies

Energy Information Administration, Petroleum Supply Monthly, May 1998.
This publication of the U.S. Department of Energy provides updated statistics on global oil supplies and imports and exports, as well as prices and a breakdown of oil products. Historical tables show changes in supplies and prices since the early 1980s.

International Energy Agency, *Oil Market Report*, July 9, 1998.
The Paris-based IEA, created in response to the 1970s' energy crises, monitors the global oil market and helps consuming countries overcome supply disruptions. The latest report concludes that the current oil glut and low prices are likely to continue until consumer nations draw down their record high oil stocks.

The Next Step

Additional information from UMI's Newspaper and Periodical Abstracts™ database

Caspian Sea

"Union Texas Expands its Caspian-Area Position," ***Oil & Gas Journal*****, April 27, 1998, p. 25.**
Union Texas Lok Batan Ltd. has purchased 75 percent of New Jersey-based BMB Oil Inc.'s 100 percent interest in a production-sharing agreement with the State Oil Co. of Azerbaijan. Union Texas's plans for the Caspian region are discussed.

Smith, Pamela Ann, "Gulf States Expand Caspian Activities," ***Middle East*****, June 1998, pp. 22–23.**
The involvement of Saudi Arabia and other gulf states in the oil and gas industries of the former Soviet republic of Kazakhstan is discussed. The oil and gas reserves of Central Asia potentially rival those of the Arab states.

Environmental Issues

"Act Now to Help Reverse Climate Change and Endorse Energy Efficiency," ***Amicus Journal*****, spring 1998, p. 8.**
The American Society of Heating, Refrigerating and Air Conditioning Engineers (ASHRAE) sets energy-efficiency standards for all commercial buildings in the United States. Because many companies complained that the new 1996 standards were too strict, ASHRAE watered them down, and the Natural Resources Defense Council is asking environmentalists to work together to have them raised again.

Dietz, Francis, " Clearing the Way for Emissions Reductions," ***Mechanical Engineering*****, February 1998, p. 36.**
European Union and other delegates prodded, scorned and shamed the U.S. delegation at the U.N. climate talks in Kyoto, Japan, to reduce U.S. emissions of greenhouse gases below 1990 levels.

"Offshore Environmental Concerns Mitigated by Onshore-Based, Extended-Reach Drilling," ***Oil & Gas Journal*** **(OGJ), May 4, 1998, pp. 118–120.**
An extended-reach well operated by Benton Oil & Gas Co., in partnership with Molino Energy Co., will be closely watched by politicians, environmentalists and industry observers. If the well is successful, it may set a precedent for offshore development without the use of new offshore platforms and reduce the risk of offshore spills.

Rowe, Duncan Graham, "Resources: Energy: Better Ways Than Burning Turn a Mix of Wood, Straw and Dung Into Gas and You Have an Efficient, Eco-Friendly Fuel," ***The Guardian*****, May 5, 1998, p. E8.**
The forest fires blazing a trail through the Indonesian rain forests and seemingly unstoppable global deforestation should make wood an unlikely contender for renewable energy. In fact, wood, straw, and animal waste, commonly known as biomass, are considered a highly desirable alternative to fossil fuels such as coal, natural gas and oil. However, the fuel must come from renewable sources, like forestry residue or short-rotation coppices (trees planted for this purpose). Unlike fossil fuels, biomass produces a neutral amount of carbon dioxide, one of the gases blamed for global warming. This means that carbon dioxide produced by burning wood is equal to the amount absorbed by the trees when they were growing.

Sharp, Linda, "Alternative-Fuel Vehicles Need Help to Clean Our Air," ***Atlanta Constitution*****, March 20, 1998, p. S2.**
The author comments on the contributions that alternative-fuel vehicles could make to cleaning up air pollution, saying that to have a major impact on the air quality of major metro areas, mass transit will have to be the major source of transportation.

Swanson, Ken, "Rebuttal: America Benefits from Ethanol Subsidies," ***Detroit News*****, April 28, 1998, p. A8.**
Instead of being chastised for their support of ethanol, Michigan Sens. Carl Levin (D) and Spencer Abraham (R) and House Speaker Newt Gingrich, R-Ga., should be commended. It's fortunate that these elected officials understand the facts about this renewable fuel and its benefits to the economy, the environment and U.S. energy security. The ethanol tax exemption is claimed by gasoline marketers, many of whom are independent, small-business owners who blend their product with ethanol. Ethanol producers benefit as well. There are 48 ethanol production plants operating in 20 states, and an increasing number of these are farmer-owned and operated co-ops that help support small towns and small businesses.

"The Wild, Wild East," ***Amicus Journal*****, spring 1997, p. 8.**
China could save more than one-third of its energy through cost-effective, energy-efficient technologies. The Natural Resources Defense Council is working with American and Chinese officials to promote this and other environmental-protection issues.

Organization of Petroleum Exporting Countries (OPEC)

Bahree, Bhushan, "Saudis Lead Plan to Form New OPEC to Boost Price of Oil by Ending Glut," The Wall Street Journal, June 29, 1998, p. A3.
Saudi Arabia, Mexico and Venezuela are negotiating at the highest government levels to create an ad hoc group of major petroleum-producing countries that would cooperate

to raise oil prices by reducing the current oil glut. The campaign, far from being a mere trial balloon, already has won converts. If successful, the new alliance would mark the most significant realignment among oil-producing nations since the formation of Organization of Petroleum Exporting Countries in 1960.

Durgin, Hillary, "OPEC Cuts Production in 'Do or Die' Situation," *Houston Chronicle*, June 26, 1998, p. C1.
Despite political entanglements at home and market-share rivalries within the Organization of Petroleum Exporting Countries, oil ministers could not overlook the economic pressures and potential chaos brought on by the lowest oil prices since 1986. For many OPEC countries, oil is the chief source of revenue. The slide in prices has meant billions of dollars in lost income, prompting budget cuts and project delays and inviting social and political unrest. The recent freefall in the price of oil to less than $12, compared with 1997's average price of $17.64 per barrel, has prompted government officials scrambling for a solution.

Hamilton, Martha M., "Oil Powers Consider Broader Group Than OPEC," *The Washington Post*, June 30, 1998, p. E3.
Major oil-producing nations are eyeing the prospect of creating a broader group than the Organization of Petroleum Exporting Countries in an effort to curtail production and keep oil prices from falling—but the plan may be destined to fail, according to some oil-industry watchers. The idea of an expanded group of producing nations that could prop up prices has been bubbling up since March. At that time, Mexico, which is not a member of OPEC, agreed with Saudi Arabia, Venezuela and other OPEC members to reduce oil production in order to boost prices. And this week Saudi Arabia's oil minister, Ali Nuaimi, was quoted in the "Middle East Economic Survey" predicting the development of an informal group of oil producers to intervene in the market, as needed. That set off a scramble by countries that depend heavily on income from oil exports to try to strengthen prices, including last week's agreement by OPEC to cut production by about 1.4 million barrels a day. But because nations outside OPEC, including Russia and Norway, are producing increasing amounts of oil, it may take more than efforts by the fading OPEC cartel to do the job.

Ibrahim, Youssef M., "OPEC Reaches New Deal to Cut Oil Production," *The New York Times*, June 25, 1998, p. D1.
Overcoming a serious political dispute that split the Iranian delegation and threatened the final agreement, OPEC announced late today that it planned to reduce its oil production by more than 1.3 million barrels a day, or nearly 5 percent, in the latest attempt to push up prices. The much-anticipated agreement was announced here at the meeting of OPEC oil ministers near the close of the commodity markets in New York today. The dispute among the Iranians had sent oil prices dropping during the day, wiping out an early rally in European markets after reports that a deal was imminent. But prices ended little changed, with crude oil for August delivery settling up 8 cents, at $14.60 a barrel, on the New York Mercantile Exchange. The Iranian split, as well as OPEC's reputation for being unable to maintain production cutbacks, left some experts questioning how long the agreement would hold. "It is a fragile agreement by its very nature, as it depends on the pledge of each country to cut its production," said Fadhil al-Chalabi, a former OPEC under secretary. He added that some countries might again be tempted to increase production if prices rise.

"Price Retreat Sends Message to Oil Producers, Analysts Say," *Houston Chronicle*, June 26, 1998, p. C3.
Crude oil futures prices retreated Thursday on the New York Mercantile Exchange, sending a message to world oil producers that pledges of deep output cuts must be fulfilled to end a supply glut and boost prices. Crude oil resumed its recent slide one day after members of the Organization of Petroleum Exporting Countries agreed in Vienna, Austria, to the third round of production cuts this year in an effort to combat falling prices. Analysts have said world oil producers must slash at least 3 million barrels a day from exports to see prices rise significantly, and the new agreement brings the total from all producers—OPEC and non-OPEC—to 3.1 million barrels, or 4.4 percent of daily consumption.

Rynecki, David, "Price of Oil up 13 Percent; OPEC May Trim Output," *USA Today*, June 23, 1998, p. B1.
Speculation that major oil exporters will agree this week to slice production lifted crude oil prices more than 13 percent Monday, the steepest gain since the Persian Gulf War in 1991. Crude oil prices rose $1.59, or 13.4 percent, to $13.43 a barrel on the New York Mercantile Exchange. The move came as ministers of the 11-member Organization of Petroleum Exporting Countries prepare to meet in Vienna Wednesday. Analysts largely expect OPEC to cut daily production, now about 28 million barrels, by 700,000 barrels to offset weak demand.

U.S. Oil Policy

Mossavar-Rahmani, Bijan, "Time Ripe to End U.S.-Iran Impasse," *Oil & Gas Journal*, Jan. 26, 1998, pp. 33–36.
The author discusses the importance of oil in U.S.-Iranian relations. He believes ignoring the importance of Iran's oil is a mistake and discusses what needs to be done to probe and challenge the existing political, legal and institutional constraints in order to put oil on the front burner.

Yardeni, Edward, "A Race Against the Calendar," *The New York Times*, Dec. 7, 1997, p. 13:
The year 2000 problem poses a serious threat that could disrupt the U.S. economy and bring about a yearlong global recession, beginning in January 2000. Such a recession could be as severe as the 1973–74 global downturn caused by the OPEC oil embargo. The 2000 problem is both trivial and overwhelming. Unless fixed, most computers, including many PC's, will produce nonsensical results or crash because "00" in the widely used two-digit year field on the computer screen will be recognized as 1900 rather than 2000. Assessing the likelihood that the year 2000 problem will set off a recession—as well as measuring its depth and duration—requires answers from government and business managers around the world.